Food and Water Security

Javid Parray · A. K. Haghi
Editors

Food and Water Security

Sustainable Solutions

Springer

Editors

Javid Parray
Department of Environmental Science
Government Degree College Tangdhar
Kupwara, Jammu and Kashmir, India

A. K. Haghi
Department of Chemistry
University of Coimbra
Coimbra, Portugal

Research Chemistry Centre Edinburgh
Edinburgh, UK

ISBN 978-3-032-04869-1 ISBN 978-3-032-04870-7 (eBook)
https://doi.org/10.1007/978-3-032-04870-7

This Springer imprint is published by the registered company Springer Nature Switzerland AG
The registered company address is: Gewerbestrasse 11, 6330 Cham, Switzerland

Introduction

This book offers a comprehensive exploration of food and water security, catering to a diverse audience ranging from students and budding engineers to seasoned professionals for sustainable development. This research-oriented book is a vital source for food, agricultural, biological, and chemical engineers, and industry experts with strong emphasis on the relationship between food and water security. This volume links theory and practice and provides comprehensive, up-to-date technical information and it delves into a wide array of crucial topics, each meticulously crafted to provide valuable insights into the complex world of food and water security for sustainable development. It addresses the paramount concern of safety in the food industry and water security, tackling challenges and opportunities in ensuring the quality and integrity of food products clean water. This research-oriented book examines the key issue of maintaining food safety and water quality as preservation methods become more varied and complex. The book also looks both at individual technologies and how they are combined to achieve the right balance of safety, quality and shelf-life for particular products.

This book considers topics in light of factors such as safety and environmental issues and serves as an invaluable resource for students, educators, researchers, and professionals engaged in the food and water industry. Whether you are an aspiring engineer, a seasoned expert, or a curious consumer, this comprehensive guide offers a deep understanding of the complexities and innovations in food processing, addressing critical challenges and charting a course towards a safer and more sustainable food future.

The chapters also cover meaningful and productive experimental findings to report hazard analysis and human health risk assessment. It reports practical achievements in food safety and water pollution control along with the technological innovation methods for better managing for continuous screening and monitoring of chemical residues in water network.

Chapter 1 of this research-oriented volume delves into the fundamental tenets of food process engineering and agriculture, underscoring the application of principles drawn from various engineering disciplines, concepts, and implications.

Chapter 2 explains in detail that there are a several obstacles and opportunities that arise in the endeavor to guarantee the food safety and water sustainability over the entire world. Challenges discussed in this chapter include contaminants (microbial, chemicals, physical) and allergen mismanagement, innovative food preservation procedures along with water safety interventions.

The aims of this chapter is to improve the everyday quality of life using technological innovations that are revolutionizing food and water sustainability and quality, including automated systems and precision agriculture towards SDGs.

Chapter 4 addresses the classification of raw materials, the requirements for raw material analysis, the fundamentals of analytical methodologies, and common techniques for food safety analysis, verification and global risks.

Chapter 5 covers state-of-the art knowledge on several key aspects of modern energy efficient drying technologies for sustainable food security along with highlighting potential application, the quality, safety and energy aspects. It provides readers with an overview of the latest applications of various novel-drying technologies in sustainable food security and processing. It clustered around major themes on drying of specific food products for safety and also the sustainable food security.

The key issues on climate change and the current challenges on water and food security presented in Chap. 6 that illustrates the fundamental impacts of climate change along with various techniques adopted for sustainability of by food processing technologies.

Chapter 7 reports the most recent technological achievements for food and water security, laying the groundwork for further study and research. It also presents cutting-edge information on new and emerging food engineering processes to improve the human health. It gives an overview on basic models, quality, safety, and technologies associated with food and water security operations today.

The book delves into key topics on sustainable agriculture: role and importance in Chap. 8 for precision farming, natural resource optimization, and real-time monitoring of agricultural systems. It examines the integration of new technologies as a pathway to reducing environmental impact, enhancing agricultural productivity, and adapting to climate variability.

This chapter is tailored for researchers, students, farmers, agribusiness professionals, and policymakers seeking advanced strategies for sustainable agricultural management as an essential guide for anyone interested in leveraging cutting-edge technology to revolutionize agriculture and secure the future of global food systems.

Chapter 9 addresses that quality management systems are crucial in each sector of the food industry, to ensure safe, quality food in a globalized world. It also covers various risk-based food safety standards to prevent health issues from occurring. The chapter also presents how the food safety initiatives of governments, intergovernmental organizations, and other interested parties are affected by the actions of the international standards systems.

Chapter 10, envisions the evolving roles of global food systems and agriculture, emphasizing innovation, adoption strategies, and potential breakthroughs. It discusses anticipated advancements in technologies for addressing future challenges in food security and water quality for sustainable development and resource management.

Chapter 11 elucidates the crucial role food and water security technical innovations in meeting the evolving demands of world population while taking into account ethical and social considerations. Furthermore, the chapter will also highlight the significance of building a sustainable and resilient food supply chain within the food and clean water-processing sector.

The final Chap. 12 provides an overview of recent technological advancements in the RNAi-based Gene Silencing Technologies as New Plant Breeding Techniques (NPBT).

It highlights cutting-edge trends that enhance efficiency and product quality and showcases successful cases of utilizing food products to maximize resource utilization.

This book serves as an invaluable resource for students, educators, researchers, and professionals engaged in the food and water industry. Whether they are an aspiring engineer, a seasoned expert, or a curious consumer, this comprehensive guide offers a deep understanding of the complexities and innovations in food processing and water safety, addressing critical challenges and charting a course towards a safer, more sustainable food future.

Srinagar, India — Javid Parray
Coimbra, Portugal — A. K. Haghi

Contents

Part I
Foundations of Food and Water Security

Chapter 1
Food and Agriculture: Concept and Implications

Wasswa Shafik, Ali Tufail, Liyanage Chandratilak De Silva, and Rosyzie Anna Awg Haji Mohd Apong

Abstract The relationship between food and agriculture is multi-dimensional and complex. Understood narrowly, agriculture is merely a means for food production. At another level, agriculture transcends its immediate role as a provider of food and plays an important part in both economic development and social change. Nowhere is the question of agriculture more central the world over than when it relates to the provision of food for the millions of underprivileged and the need for global food security. For a majority of low-income nations with large agriculturally dominant economies, agriculture is also a matter of vital concern as a major source of employment and income, providing economic opportunities for food, livelihood, and employment. Though history is replete with instances where civilizations grew and ultimately collapsed when agriculture was no longer sustainable the way it was practiced, modern agriculture faces a different range of challenges, including growth and efficiency in food production, diversity of food production systems and diets, market opportunities, and distribution of economic rewards and, crucially, regulation and governance. Production systems, as they have evolved historically, straddle a continuum that draws from effective resource management and optimum location of food production systems and infrastructures on one end and the maximizing of food production through ecological intensification, driven by the use of chemical and biological technologies and inputs, on the other. At another level are the emerging cuisine sectors, some of which are driven by cultural and dietary reasons, while others by health and ethics considerations. The challenges faced by developers today are thus multi-dimensional and require eco-socioeconomic approaches and solutions.

Keywords Food systems · Agriculture · Sustainable farming · Food security · Agricultural practices · Climate change · Resource management · Food supply chain

W. Shafik (✉)
Dig Connectivity Research Laboratory (DCRLab), Kampala, Uganda
e-mail: wasswashafik@ieee.org

W. Shafik · A. Tufail · L. C. De Silva · R. A. A. H. M. Apong
School of Digital Science, Universiti Brunei Darussalam, Tungku Link, Gadong, Brunei

J. Parray and A. K. Haghi (eds.), *Food and Water Security*,
https://doi.org/10.1007/978-3-032-04870-7_1

1.1 Introduction

The connexion between food and agriculture is multi-dimensional and complex. Understood narrowly, agriculture is merely a means for food production. At another level, agriculture transcends its immediate role as a provider of food and plays an important part in both economic development and social change (Nawaz et al., 2024). Nowhere is the question of agriculture more central the world over than when it relates to the provision of food for the millions of underprivileged and the need for global food security. For a majority of low-income nations with large agriculturally dominant economies, agriculture is also a matter of vital concern as a major source of employment and income, providing economic opportunities for food, livelihood, and employment (Jadhav et al., 2024).

Though history is replete with instances where civilizations grew and ultimately collapsed when agriculture was no longer sustainable the way it was practiced, modern agriculture faces a different range of challenges, including growth and efficiency in food production, diversity of food production systems and diets, market opportunities, and distribution of economic rewards and, crucially, regulation and governance (Liu et al., 2024). Production systems, as they have evolved historically, straddle a continuum that draws from effective resource management and optimum location of food production systems and infrastructures on one end and the maximizing of food production through ecological intensification, driven by the use of chemical and biological technologies and inputs, on the other (Rozenstein et al., 2024). At another level are the emerging cuisine sectors, some of which are driven by cultural and dietary reasons, while others by health and ethics considerations. The challenges faced by developers today are thus multi-dimensional and require eco-socioeconomic approaches and solutions. Overall, therefore, an understanding of food and agriculture today needs to be based on integrated ecological, social, and economic understandings (Pascutti et al., 2024). Only then would it be possible to develop solutions that are sustainable in a holistic sense.

1.1.1 Definition and Scope

Agriculture is generally viewed as including both primary production and the activities directly related to products as they leave the farm. This definition has the advantage of being largely recognized all over the world and reflects the multi-sectoral characteristics of agriculture, sometimes called the rural world. Indeed, primary farming is usually distinguished from agri-food chain-related activities (Gera & Jain, 2024). These so-called related activities are variously defined in order to suit major purposes such as statistical scope, public policy programs, and academic research. As far as traditional farming is generally taken into account, the availability of food is more based on food self-sufficiency linked to everyone's capability to produce and process food at the household level. It tends to cut the commercialization sector described above and is also called secondary agriculture (Li et al., 2024).

Every time the concept of agriculture is considered and investigated, a few terms that are highly involved, too, are food security, sustainable development, and agricultural productivity. Livelihood and ensuring food availability in any society are among the most important collective actions of people. Meeting the growing demand for food, energy, and water that is economically feasible, environmentally sound, and culturally acceptable is the finality of all societies living on earth while maintaining the balance of local, national, and global demands and the availability of usable resources (Vishnoi & Goel, 2024). These preoccupations must be transversal to all activities that aid in the production and distribution of food, i.e., in all the sectors constitutive of the agri-food world.

1.1.2 Historical Development

Agriculture is considered to be as old as human civilization. It has contributed immensely to the growth of human civilizations and the societal changes required for them. Agriculture in ancient times established irrigation systems and land requirements that gradually forced primitive people to become semi-nomadic and then fully nomadic. Slowly, as the population grew, agriculture became widespread due to the cultivation of cereals and pulse crops and the domestication of animals (Haydar et al., 2024). Agriculture was the foremost activity performed by hunter-gatherer tribes long ago, as we see in ancient texts and histories of Central Asia and the Middle East. Food security was taken into consideration since agriculture provided a reliable source of food.

As time passed, technological inventions such as plows, tractors, drills, harvesters, and irrigation systems brought greater benefits. Not only these, but small innovations in sowing techniques and seed preparations also brought significant changes in the development of agriculture. In some countries, land laws and inheritance laws led to the bundling of initially large tracts into smaller one-family farms (Obaid et al., 2024). The extension of cultivation was manifested in many ways as new cultures were introduced, especially with the introduction of plantation economies in various countries. The promotion of research and extension work was important in many developing countries. As a result, this brought a lot of development in agricultural policies, such as the Corn Law and the differential price concept in the commodity variety artifacts. Governments also tried to stabilize crops earlier under local rulers and mercantilism to assure their people against famine. In India, administrations allowed a sliding scale of duty on wheat imports as droughts appeared (Xun et al., 2024). Thus, agriculture today has grown from many such ancient policies and strategies that have shaped the basis for modern agricultural countries.

1.1.3 Key Concepts in Agriculture

Of the various origins of human civilization, growing food was the first primary avocation by which various ethnic groups around the world have achieved self-dependence and self-reliance. Agriculture supplied them with soft and hard services that played a vital role in securing a niche close to or in the vicinity of forests, as these provided livestock fodder, fuelwood, and environmentally friendly materials that together made life sustainable for agriculture (Obaid et al., 2024). The industrial and neo-industrial models of agriculture may be able to supply these services and show high technological advancement, yet there is a trade-off that comes with such "advancement." In other words, such farming systems may not be sustainable. Consequently, the millennium has witnessed an interest in sustainable agriculture models and agroecological theories as an alternative principle of agriculture (Haydar et al., 2024). The main principle that is common between sustainable agriculture and agroecology is the same: minimizing energy inputs in favor of greater reliance on natural and human resources.

Agroecologists have formally divided this guiding principle into four sub-principles that guide all agricultural management decisions. These sub-principles are recycling of resources, regeneration rather than depletion of agricultural and surrounding resources, dependence on internal agroecological dynamics rather than purchased goods, and internal self-regulation rather than external regulation. These principles are not necessarily what evolution tells us about ecosystem attributes that might allow the resilient provision of services (Vishnoi & Goel, 2024). Agriculture is essentially part of the terrestrial biosphere, and it is fertile soil with which we grow food; the more fertile the soil is, the more resilient the agroecosystem emerges from incorporating these principles. This is service research by ecology as well as a synthesis that incorporates diversity on the one hand and farmers' knowledge systems on the other as its two most fundamental and historic guidelines (Li et al., 2024). Biodiversity is service research's insurance policy and one that does not cost.

1.1.4 Sustainable Agriculture

Today, interest in and demand for "sustainable agriculture" are strong. The concept is sorely in need of careful definition, and the term justifies a wide variety of understandings, as sampled in Fig. 1.1. Many people take sustainability to refer to practices that are, over some unspecified period, able to provide a region's food without depleting its resources or harming its environment. This idea has proved to be a compelling one. Clearly, we want to be able to feed ourselves and future generations without compromising their ability to do the same (Haydar et al., 2024). In an age when we begin to realize how badly off we have managed to get our environment and how constrained the world's food supplies are, the allure of "sustainability" is not hard to understand.

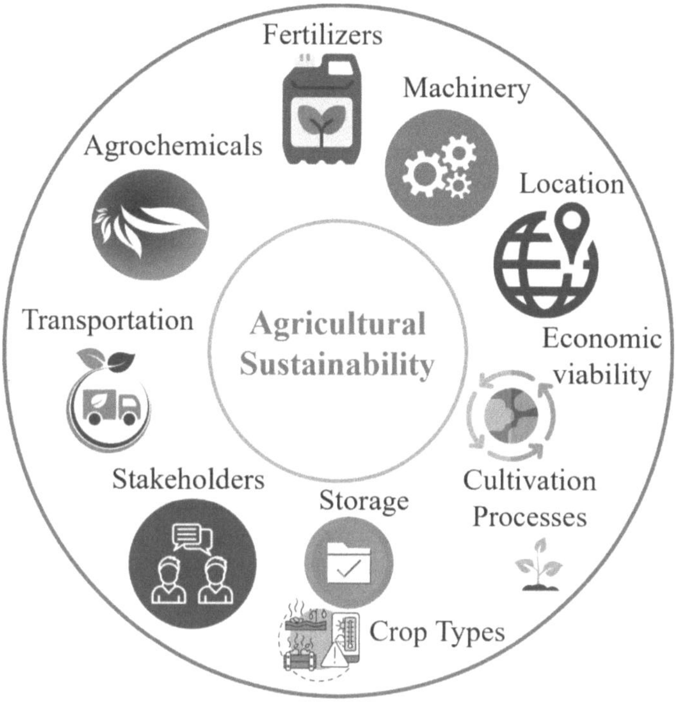

Fig. 1.1 Sustainable agriculture

Advocates of sustainable agriculture promote a variety of practices that are thought to contribute to sustainability. They promote soil health and reduce chemical inputs. They maintain or enhance biodiversity. They protect the quality of the air and water. They conserve resources such as water and soil. Many sustainable agricultural practices are less "intensive" than increasingly dominant industrial practices but do not have to produce less (Obaid et al., 2024). Sustainable farming is often dismissed as more expensive and less "efficient" than industrial practices, but many advocates of sustainable agriculture say that this is because industrialization makes up or overlooks many costs that never appear on the farm balance sheet. There is a large and growing interest in the idea of the triple bottom line in business management (Xun et al., 2024).

However, a truly complete environmental accounting must include future costs. The practices that are often called sustainable do not necessarily produce less in the short run, but they may require more labor at certain periods. They may not be as resistant to certain pests and diseases or as adaptable to the demands of modern agribusiness or modern life. They require different skills, different infrastructure, and different markets. Numerous policy supports are necessary for the promotion of sustainable agriculture, but ultimately, a grassroots surge of interest and support from the non-commercial sector is probably the strongest driver (Tokumaru et al., 2024).

The technological limitations of sustainable agriculture include poor information and a lack of communication between different fields. Challenges to sustainable agriculture are the existence of other powerful, determined influences. Importantly, the very diversity of goals expressed in sustainable proponents' proposals, possibly a function of just how open and elastic the concept remains, makes the widespread implementation of deeply sustainable agriculture unlikely when the larger food system in which farms are situated operates so very differently (Kayastha et al., 2024). This makes the cultivation of a large and enthusiastic consumer base that appreciates and demands sustainable goods especially important in the fight for a more sustainable food system. It is also enough to make the hard-core sustainability advocates despair of convincing their compatriots in the food world of the proper order of things.

In sum, explain the intersectional relationships of healthy agriculture, viable ecology, and a just society in the world, as well as the different possibilities for practicing those intersections on an individual level. As people start to be more and more aware of these concepts, the "how" of sustainable agriculture becomes more difficult to answer. It is clear, however, that sustainable agriculture can be practiced by the work of individual, like-minded farmers committed to caring for their land through the profitable production of healthy food (Shafik, 2024b).

1.1.5 Agroecology

Agroecology is the science of applying ecological principles to the study, design, and management of sustainable agroecosystems. It is an approach to farming that gives the concept of food-producing ecosystems more prominence. Agroecology is based on applying the principles and practices of ecology to the design and management of sustainable food systems. Agroecology seeks to optimize the connections between parts of the food system, which include seeds, plants, insects, animals, soil, water management, and waste cycles, to optimize the interactions and reduce the need for external inputs, create synergies and gradualism, substitute biological inputs for chemical inputs, and bring about beneficial arthropod extension, and develop a biologically efficient, productive system as a foundation on which to establish economic viability, including the stewardship of water and soil (Liu et al., 2024).

Agroecological techniques include the rotation of different crops, particularly nitrogen-fixing species, to reduce pests and diseases and build soil nutrients, rather than the continuous growing of just a single crop. Regularly rotating crop species in a field is critical to managing pests, particularly crops that belong to different plant families, because most pests are specialists that will only feed on plants belonging to one related group. Another technique is intercropping, which is growing two or more crops together in proximity (Rozenstein et al., 2024). This practice can help agriculture be more resilient and productive, as a diverse array of crops tends to share resource requirements. Agroecology can also include keeping animals such as chickens or ducks free-ranging among croplands. This increases fertilizer inputs from the animals, as well as eating many insects. Agroecology is about the relationship between farming, nature, and food. It also encompasses aspects related to climate change and food security (Jadhav et al., 2024). Peasant-led agroecology is also rooted in food sovereignty, which is the right of people to food and to produce food.

1.1.6 Food Production Systems

Several food production systems are practiced in different sectors of agriculture. Farmers involved in conventional farming follow a high-input path, which relies heavily on synthetic pesticides, chemical agents, and other similar inputs. Conventional systems favor monoculture practices and have shown impressive crop yields, though with some question marks on their long-term sustainability (Pascutti et al., 2024). The shift from the conventional approach has brought in several systems with a degree of alternative thinking, most notably in organic farming systems. The primary focus in organic systems is to promote ecological, economic, and social sustainability and optimize long-term biological productivity and biological sustainability in suitable agroecosystems (Liu et al., 2024).

Conventional agricultural practice has its pros and cons: high productivity levels to feed the new generation of people, especially in times of chronic food shortage, displaced people who require food and nutrition, emerging health issues with a possible cure from crop-based chemicals, and the creation of the latest biotechnologies to feed the next generation. There is also the need to look at the reasonable and minimal use of chemicals for sustainable development. Other ecological issues need to be studied from an environmental view with no harm and pollution, safe for farmers, and an industrial standpoint, high profitability. Organic farming is a production system that avoids or largely excludes the use of synthetic inputs such as fertilizers, pesticides, hormones, and feed additives to the maximum extent feasible, as governed by external and internal influences (Pascutti et al., 2024). Organic farming is not just a philosophy; it is a well-demonstrated scientific and agricultural practice of avoiding harmful synthetic chemicals and developing instead eco-friendly biological systems using all plant parts as well as waste materials. The underlying principle of organic agriculture is to allow environmental processes to thrive in harmony with agricultural systems. Organic farming is a fast-growing consumer movement in developed countries and is practiced over several concepts to a various extent in others across the world. There is an emerging holistic set of markets for realizing values attached to minimally and truly organic produce, which will give them a margin in earnings that will be passed on to the producer (Vishnoi & Goel, 2024). The overall assessment is that one has to carefully decide on food and its production, particularly which has a direct bearing on the environment, for the long term by avoiding pollution of soil and water. The knowledge of the production systems and related services has to be strengthened for sustainable farming systems. The complexity surrounding agricultural systems has implications for food supply availability and environmental usage. Such complexity is best viewed along a continuum of food production systems (Jadhav et al., 2024; Rozenstein et al., 2024). Understanding food production is complex because there are many routes to producing a foodstuff for the consumer with varying impacts.

1.1.7 Conventional Agriculture

Conventional agriculture has been the dominant mode of providing food in many regions. In this subsection, we present the practical features of conventional agriculture, the economic rationale behind the emergence of these practices, the main social features, and the link with the process of globalization. Concerning the technological aspects, we describe the most important trends and practices employed by this model. The substitution of capital in the form of machinery and mineral fertilizers in place of increased labor has been the driving force behind mechanization and the use of agrochemicals (Gera & Jain, 2024). Since World War II, the area farmed using conventional mechanized agriculture has multiplied dramatically.

The main features of conventional farming are high yields per unit of cultivated surface, as well as low labor inputs, which lead to much higher income compared to extensive agriculture. These have provided the economic rationale and stimulus for large investments in agricultural machinery, thus driving investments in agrochemicals. The consecutive use of agrochemicals guarantees that the most limiting factors do not determine the final yield but only the cultivated surface (Haydar et al., 2024). For input efficiency and strong mechanization, the plants are kept on a sole vertical axis. Pesticides, as well as mineral fertilizers, play a very substantial role in the international agricultural economy. Economic trends can help one to understand the scale of this dependence more clearly.

1.1.8 Organic Farming

Organic farming is considered to be a sustainable alternative to conventional agriculture using different agricultural aspects, as illustrated in Fig. 1.2. It is rule-based and aims at the utilization of ecological concepts and the least possible use of energy and technical means. The major principles of organic farming are

(1) ecological balance;
(2) biodiversity;
(3) the precautionary principle, the basis of which is the ecological view of the agricultural landscape; and
(4) health, culminating in the concept of clean.

"Health" here means the absence of pollutants and residues in the soil and food (organic farming is often described as "clean" or "contamination-free" agriculture). Everywhere, this form of agriculture has to comply with national or international

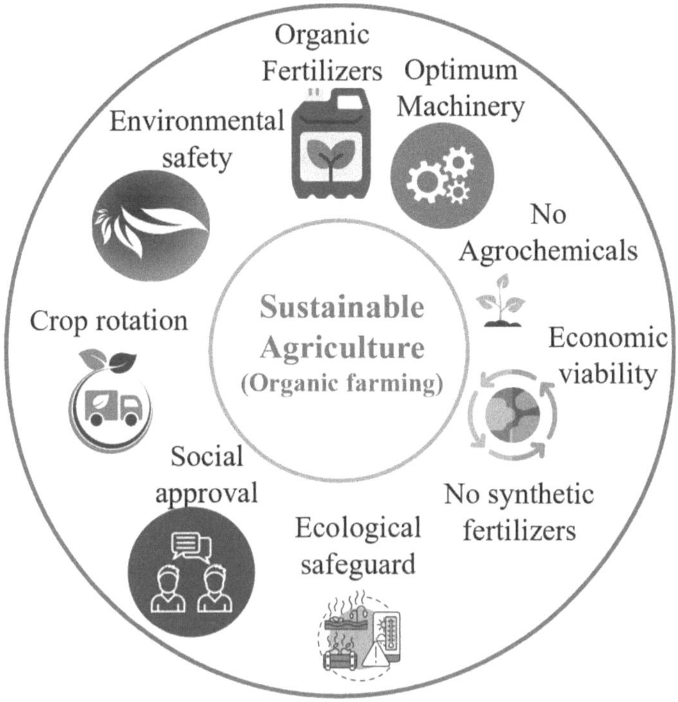

Fig. 1.2 Organic farming

regulations based on the rules for organic farming and take part in state-controlled certification procedures to be allowed to use the label "organic" for selling products (Obaid et al., 2024). This certification is associated with certain costs and obligations, e.g., traceability of products and documentation.

Although there are two main European and American certification standards, there are many others. The goal of organic farming is to create integrated, humane, environmentally, and economically sustainable food systems. Organic practices are designed around four health-of-the-earth principles: ecological, economic, social, and humane. Organic farmers are facilitators in making these broad, holistic connections that people make as they eat. Most of the food safety problems are related to industrial agriculture, not organic farming (Kayastha et al., 2024). There are many challenges for organic farmers in the production of food in their local area and selling it to the market at a competitive price. A study on challenges in organic farming uses a variety of qualitative and quantitative data to document the key challenges facing organic farmers, challenges related to pollution and contamination, challenges associated with marketing and regulation, as well as challenges associated with the operation and maintenance of the farm (Mohammadi et al., 2024). One of the primary challenges regarding organic crop production identified was the competition for markets. Certified organic and non-certified organic producers often have the same or similar end markets. Organic agriculture leads to the production of safe, nutritious, uncontaminated food or raw materials and non-neutral or even beneficial effects on the environment, compared with using non-renewable resources and eliminating pollution. The use of residues from organic farming also contributes to increasing food security and improving rural community health. Therefore, promoting organic farming is part of the World Food Declaration (Saikanth et al., 2024; Shafik et al., 2023). Organic farming is increasingly looked upon as a sustainable alternative to conventional agriculture.

1.2 Challenges in Agriculture

Agriculture today faces significant challenges. What is grown and how it is grown, from a choice of seed and livestock breeds to water and energy inputs, strongly affects how it will contribute to nutrition and human health, economic development, or ecosystem integrity. Agriculture itself is affected by changes in climate. In turn, the food system can be a major driver of climate change, usually in a negative direction, because it contributes to the buildup of greenhouse gases in the atmosphere (Kalinaki et al., 2023; Shafik, 2024a; Shafik et al., 2024a). The Earth's climate has always been variable, within and among different regions and scales of time. While a growing majority of the scientific community now accepts that global warming is taking place as a result of human activities, regional impacts, and associated changes in weather patterns are both more uncertain and make a larger difference to people's lives (Shafik & Kalinaki, 2024; Shafik et al., 2024b).

Not all changes in the climate are necessarily negative for agriculture. However, current indications are that severe weather events will become an increased risk, as are shifts in precipitation patterns, such as altered frequency of droughts and floods. These not only affect agricultural productivity now but also have, through time, in some cases, led to the wholesale abandonment of successful water management techniques that have been developed over centuries (Zain et al., 2024). One of the central components of the agriculture sector is to ensure that food and other agricultural products are made available to meet the needs of a growing population in a socioeconomically and environmentally sustainable way. This ongoing production must come in the context of decreasing availability of resources, specifically land and water. Such ongoing innovation occurs within the biophysical, economic, and social limitations of specific sites, economies, and cultures. Environmental issues include the need to conserve and restore the resource base, maintaining biodiversity to contribute to resilience and innovation, and the need to minimize detriments to ecosystems (Jose et al., 2024).

Reducing biodiversity by concentrating land uses reduces the insurance of species against natural shocks, increases capital intensity, and impedes feedback, increases entropy and the likelihood of system breakdown. New policies and market forces are needed that foster these changes, rather than those that maintain systems that are optimized for the current, not the future. The complexity of these interacting forces and the dynamic nature of change mean that ongoing "solutions" are not likely to be simple and will have to be adaptive (Yuniwati et al., 2024). This overview of changes in agriculture has shown that the inevitable rise in the production of commodities that started with the Neolithic Agricultural Revolution may be set for its next seismic shift.

The current drivers will ensure that, in the next few decades, the agricultural production paradigms of the last century will continue to be challenged by food and agricultural systems that are much more than just agricultural systems. Indeed, they are best thought of as "agri-food systems" in which crop and livestock production is but one part of a food industry among many, each with a distinct function but all with the endpoint of delivering food to the consumer. Thus, food "preparation" is a much more diverse and much larger industry than food production (Katti et al., 2024; Moreira et al., 2024). Furthermore, the drivers dictate that the agricultural systems will need to grapple with some fundamental technological and social changes to address the scale of change required, especially in the intensity of, and the level of investment in, the use of resources such as crop breeding, water, nutrients, and other inputs including agrochemicals and energy (Pszczółkowski & Sawicka, 2024).

1.2.1 Climate Change and Agriculture

Climate change is a threat multiplier with far-reaching implications for global agriculture and food systems. It is expected that climate change will negatively affect the yields of the major crops. This is in part due to the amplifying climate responses of crops themselves, especially for suboptimal temperatures and high latitudes. Increases in carbon dioxide levels can have direct benefits to crops, albeit dubious, but not enough to mitigate other adverse impacts, including the impacts on pests (Jadhav et al., 2024). The degree to which these impacts are materialized depends on various factors such as the emission pathway considered and, as a consequence, temperature rise as well as precipitation changes. Furthermore, the use of water for irrigation will increase, which can also affect the sustainability of water resources. The welfare impacts of climate change are uneven across the globe, with bigger threats to rural populations of the tropics and the subtropics (Liu et al., 2024; Rozenstein et al., 2024). All in all, the socioeconomic impacts of climate change are quite uncertain, not least because there are further multifaceted linkages at the ecosystem level. Although investment in infrastructure and institutions can increase the resilience of agricultural strategies to climate change, societies will ultimately have to invest in reducing greenhouse gas emissions if significant negative impacts are to be avoided. The negative impacts of climate change must be incorporated into policy-making processes (Pascutti et al., 2024).

The emission of greenhouse gases causes changes in the climate, which in turn mediate the availability and quality of land, water, and biodiversity. In agricultural economies, a healthy natural environment represents the basis of food. Two-thirds of the greenhouse gases emitted into the atmosphere come from the use of fossil fuels through the application of fertilizers, raising livestock, manure management, and changes in land use such as deforestation. Because many of these gases are produced as byproducts of agricultural practices under the management control of individual farms, whereas clean energy technologies could provide alternative emission pathways, these reductions are related to energy demand (Gera & Jain, 2024; Li et al., 2024). Many argue that the mitigation potential is high in this area. Greenhouse gas emissions from food contribute to more than 20% of the global emissions of these gases, and 26% of the present agriculture-related emissions are due to deforestation. The development of countries will take place under curtailed resources, of which environmentally clean energy will be one at the core. Since developing countries currently have low greenhouse gas emissions from their food systems, but at the same time have high expected growth of income, their agricultural strategies should aim at developing along cleaner development pathways (Haydar et al., 2024; Vishnoi & Goel, 2024). If this is the case, agricultural growth and climate change are most likely to interact.

Agriculture has always been regarded from a multifunctional perspective, and many of the policies and strategies to increase agricultural income have focused mostly on the economic efficiency aspects. However, given the need to address other arrangements to ensure society's food security, strict consideration must be given to the singular nature and resilience of agriculture as compared to other sectors of economic development. It is recognized that agriculture will have to meet the growing demands for food, feed, fodder, fiber, and bioenergy while reducing or conserving natural resources, decreasing rural poverty, supporting a diversified rural economy, and abating greenhouse gases (Obaid et al., 2024; Xun et al., 2024). In the process, sustainable management of land, water, and biodiversity resources will have to be given equal attention.

1.2.2 Food Security

Food security—access to adequate, nourishing, and safe food for all at all times—is a fundamental human right and a critical indicator of a just society. Food security is something that we take for granted in the global North. Yet, in 2020, it was estimated that 9.9% of the US population and 8.7% of the Canadian population experienced food insecurity. Internationally, it is estimated that nearly 9% of people globally continuously face food insecurity (Tokumaru et al., 2024). The COVID-19 pandemic led to a doubling of acute food insecurity. Food security encompasses three areas: availability (sufficient, reliable, and regular sources of food), access (having financial or other means to consistently obtain food), and utilization (knowledge to choose nutritionally adequate and safe food, consumption and hygienic practices, and care to provide for adequate health and growth) (Kayaṣtha et al., 2024).

Despite scientific advancements such as increased crop and livestock yields, those who suffer from food insecurity are primarily in the global South, and huge disparities in food availability and distribution exist. The primary causes of food insecurity include poverty, social injustices such as gender-based inequality, and a lack of control over their local environment. For example, in Bangladesh, many families are considered food insecure as a result of hazardous waste dumping. As a result, rice yields decrease, and water becomes contaminated with hazardous chemicals. Environmental degradation continues to exacerbate food insecurity (Mohammadi et al., 2024). Agricultural practices that increase soil degradation can decrease agricultural productivity. Over the last 300 years, 2.5 billion people have faced a net loss of their agricultural land. Paid employment does not guarantee enough income for individuals in affected areas. A significant percentage of malnourished children worldwide live in poverty-stricken families. A notable portion of the world's population lived on less than US $1 a day; the levels of extreme poverty continue to rise in territories of Sub-Saharan Africa. In an environmental context, scarcity of water in regions with a high reliance on rain-fed agriculture has profound effects on food security; one of the key impacts of global climate change is the growing number of areas affected by droughts (Gera & Jain, 2024; Saikanth et al., 2024). Measures have been recommended on how to balance the supply and demand of water. It has been found that through integrated management and development of raw water supply on fruit trees and commercial crops, southern Africa yields may increase significantly. Moreover, rehabilitation of degraded or destroyed watersheds will concurrently improve the supply of fresh water, or where there is disruptive flooding, regulation, and use as an alternative to the construction of large dam schemes, as presented in Table 1.1 (Jadhav et al., 2024; Liu et al., 2024).

1.3 Implications of Agriculture on Society

This is still the single largest employer in the world. In general, developing countries are agriculture-based societies in which the agricultural sector accounts for almost 70% or more of the labor force. Usually, in both developed and developing countries, the agricultural sector contributes between about 20 and 50% of the GDP. In countries like the United States, agricultural output accounts for less than 4% of the national economy. However, in developing countries, output generated by agriculture accounts for a significant portion of the national economy, ranging from about 15% to more than 30%, and even up to 60% or more in some of the least developed countries (Tokumaru et al., 2024). Agriculture has a significant influence on the economy of developing nations in terms of foreign exchange. Many of the developing nations are agro-dependent and have a strong dependence on agriculture to generate foreign exchange. Uganda receives the majority of its revenues through the export of coffee, tobacco, tea, and cotton. Artisan mining in the country has also been found to be efficient for balancing Uganda's current account by generating foreign exchange (Kayastha et al., 2024).

Table 1.1 The complex effects of agricultural challenges on both human societies and natural ecosystems

Challenge	Description	Impact on society	Impact on nature	References
Post-harvest losses	Inefficiencies in storage, processing, and transport	Reduced food availability and economic losses	Waste of natural resources used in production, contributing to emissions	Jadhav et al. (2024)
Land use conflicts	Competition between agriculture, urbanization, and conservation	Strain on food production and rural livelihoods	Habitat fragmentation, loss of arable land, and decreased wildlife corridors	Liu et al. (2024)
Dependence on chemical fertilizers and pesticides	Over-reliance on synthetic inputs	Health risks for farmers and consumers due to chemical residues	Soil and water contamination, disruption of natural pest-predator balances	Rozenstein et al. (2024)
Deforestation	Clearing of forests for agricultural land	Affects local communities and indigenous populations	Habitat destruction, carbon emissions, loss of biodiversity	Pascutti et al. (2024)
Economic pressures on small farmers	High costs of inputs and market competition	Financial stress and poverty for smallholder farmers	May lead to land abandonment or unsustainable farming practices	Gera and Jain (2024)
Loss of biodiversity	Reduction in plant and animal species due to monocultures	Reduced resilience of food systems, limited crop varieties	Decreased ecosystem resilience, reduced pollination, and habitat destruction	Li et al. (2024)
Pest and disease outbreaks	Increased pests and plant diseases affecting crops	Crop loss leads to economic loss for farmers, food shortages	Increased pesticide use, which can harm non-target species and soil health	Vishnoi and Goel (2024)
Water scarcity	Limited freshwater resources for irrigation and farming	Food production declines, affecting food availability and costs	Drying up of rivers and lakes, habitat loss, and ecosystem imbalances	Haydar et al. (2024)

(continued)

Table 1.1 (continued)

Challenge	Description	Impact on society	Impact on nature	References
Soil degradation	Decline in soil fertility due to erosion, overuse, and contamination	Decreased agricultural productivity, economic strain on farmers	Loss of soil carbon, reduction in biodiversity, impaired water cycles	Obaid et al. (2024)
Climate change	Increasing temperatures, erratic rainfall, and extreme weather events	Reduced crop yields lead to food insecurity and higher food prices	Soil degradation, water scarcity, loss of biodiversity	Xun et al. (2024)

1.3.1 Economic and Nutritional Perspective

Agriculture Trade Rwanda's agricultural discrepancies are the gap between the consumption of perishable food products and the ability to supply. The gap is filled through food crop imports, which cost $100 million in 2007; this is equivalent to 18% of total imports. Agricultural trade is very important for aggregate economic welfare. Not only does it have a significant effect on the level of income and employment, but it also has a major impact on the balance of payments in countries that are experiencing trade surpluses. Agricultural trade is very important for the world's poorest countries, for whom agriculture still dominates exports. Agricultural trade is an important part of the overall global trade market (Mohammadi et al., 2024). Global food company sales in 2016 were $7.8 trillion; sales in 2017 are predicted to rise to $9.5 trillion. Leading global food and beverage companies include major industry players. These sales figures help illustrate the significant financial power of the agriculture and food sectors. In the US, the agriculture and food sector provides around 21 million jobs. Food manufacturing creates about 15.9 million jobs in the country. The contribution of agriculture and food to the US economy is substantial. However, the key role these industries play in employment is in rural areas, with the more industrialized urban areas employing relatively fewer people (Nalluri & Karri, 2024; Saikanth et al., 2024). This indicates that rural areas house the US agricultural industry, which is important from an economic and nutritional perspective.

1.3.2 Economic Impact

While the national shares of GDP provided by the agriculture and food sectors may be declining, their economic contribution to national economies cannot be underestimated. In developing countries, agriculture and agribusiness provide the most jobs for people as labor, self-employed individuals, or employers in the informal sector of their economies. Governments appreciate that exports from their agricultural and agribusiness sectors can deliver more foreign exchange than most other products apart from minerals (Dai et al., 2024). Agriculture and agribusiness are valuable in generating gross domestic product for national economies, while companies working in the manufacturing and services sectors also profit from key inputs, outputs, and shared marketing and distribution services (Sharma et al., 2024).

Whereas the focus in this section is on the economic importance of agriculture, it must be noted that agriculture is also important for many other reasons. The inadequacy of agricultural policies, services, and infrastructure still results in an overall agriculture sector that is a significant drag on overall economic performance. The improved performance of the agriculture sector, particularly for smallholder farming families and emerging agribusinesses, can drive the overall economic and national growth of the nation (Jose et al., 2024). Economic, social, and community performance is linked from the farm right through to the final value-added point of sale, where the food, beverage, or other agricultural product is sold. In reviewing the economic importance of agriculture and agri-food, it needs to be said at the outset that its main focus is on the potential it can generate for transforming developing economies from developing to a developed state over the long haul (Zain et al., 2024). Farming is the best way to start the economic upliftment of the greatest proportion of a nation's citizens.

1.3.3 Social Impact

Agriculture, as both a practice and concept, has profound implications for its intersection with social issues, equity, social relations, and culture. Agriculture creates both wealth and poverty, intervening directly in the object of social equity through the control of land and resources. The way food is produced is tied to markets, livelihoods, and, ultimately, the health and well-being of the urban and rural populations that it supports (Katti et al., 2024; Yuniwati et al., 2024). As such, agricultural practices and policies directly influence social equity, children's education, the rights of laborers, gender roles, and access to food. Agriculture can tie people into a "poverty trap" or offer entry points into upward mobility. Additionally, agricultural practices impact local communities and society by eroding biocultural diversity and traditions from the local and global contexts (Moreira et al., 2024). Culturally, foodstuffs and food production play a key part in constructing culture and sentiments of place.

Sustainable agriculture can lay the foundation for a wider public good in society by promoting and realizing social well-being on a number of fronts. The global trend toward the concentration of productive resources such as land, forests, water, genetic materials, and marine ecosystems has direct implications for the livelihoods and development of the majority of rural and indigenous communities around the world (Pszczółkowski & Sawicka, 2024). Ecological foods and bio-products play a diverse symbolic role around the world in terms of alternative social markets, attempts to carve a viable niche for family farmers, commitments to environmental and human health, types of exclusive goods or commodities, expressions of regional economic development, expressions of cultural revival, and territorial "re-rooting" (Moreira et al., 2024). Agriculture can contribute to enhancing the environment, people's health, and well-being on the local and regional level and make a modest, albeit significant, contribution to mitigating climate change through the reduction of greenhouse gas emissions. Policymakers must grapple with enacting policies that both enhance the uptake of sustainable agriculture and help the international economy deal justly with groups unable or unwilling to transition away from high-input agriculture (Yuniwati et al., 2024).

1.4 Technological Innovations in Agriculture

Advances in technological innovations have significantly changed the nature of agricultural practices by increasing productivity and efficiency, providing market access, improving the quality of agricultural products, and ensuring a more environmentally friendly, sustainable food production system. The adoption of information and communication technologies in agriculture has gained substantial potential (Moreira et al., 2024). Precision agriculture, known as prescription farming or precision farming, is one of the widely adopted forms of information technology in domains associated with agriculture. Precision agriculture links inputs such as water, nutrients, and pesticides with the spatial requirements of crops, thereby making them responsive to these inputs when and where necessary (Katti et al., 2024). Data intensity will characterize the future of farming, necessitating proper and accurate management of this immense volume of crop-monitoring data (Fig. 1.3).

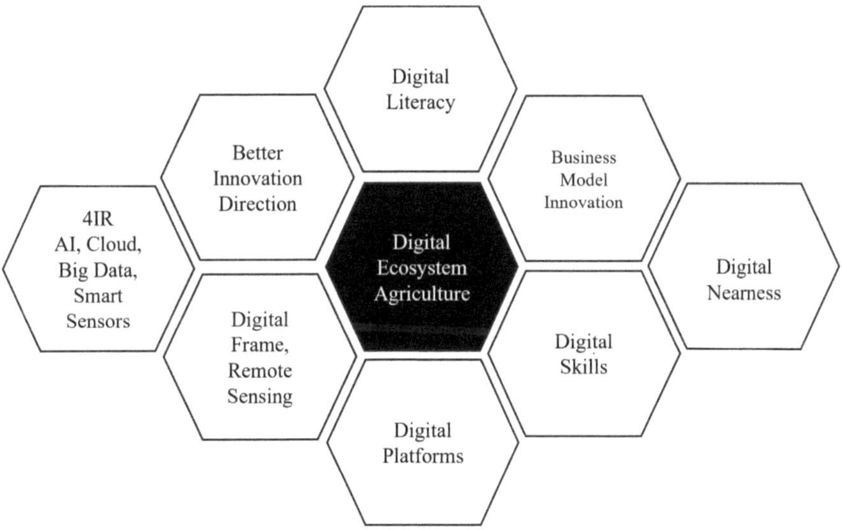

Fig. 1.3 Digital ecosystem agriculture

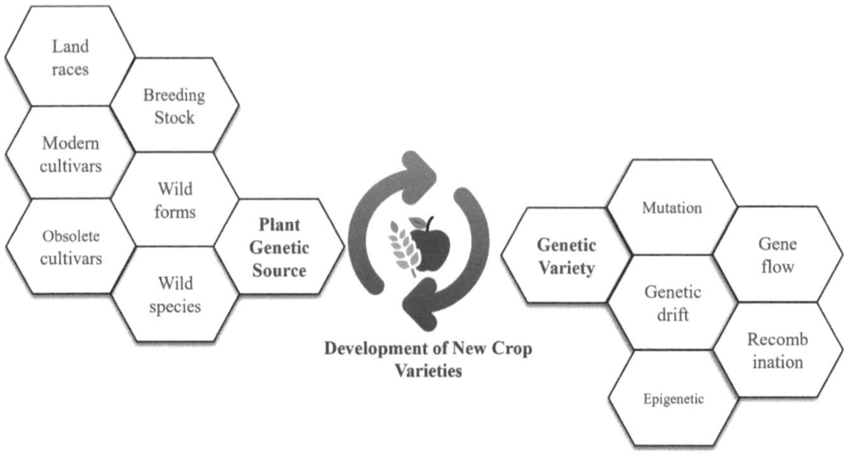

Fig. 1.4 Development of new crop varieties

Data-driven farm management approaches that place data in a central position and use decision-making tools to evaluate and act upon this data are often termed Decision Agriculture and digitalized ecosystem systems, as illustrated in Fig. 1.4. Thus, these new technologies initiate the change from precision farming as a petroleum-based and input-driven technology to data-driven agriculture with a view to using information within the field boundary on a yearly, weekly, or daily basis (Moreira et al., 2024). This, in turn, enables managers to reduce resource usage, optimize yields, increase profits, and make more efficient and sustainable use of the environment. The technological innovations, however, could further translate to low or no cost to the user base due to the digital divide. If the technology becomes too expensive for smaller interest groups to afford, it could simply alter the state of food production (Yuniwati et al., 2024; Zain et al., 2024). Simply put, the technological advances will or could have the impact of consolidating food supply in larger agro-companies, as they are most likely to afford and utilize them. Research and development will spearhead the development of new and existing technologies. Let us take a hypothetical look at biotechnology and how it can completely revolutionize agriculture (Jose et al., 2024). In fact, for some time now, there have been early signs of such developments, particularly regarding the increasing efficacy of crop production.

1.4.1 Precision Agriculture

Other approaches to farming view it as a practice that works best when it is left alone. Precision agriculture is altogether different: it is an approach that uses a combination of GPS, sensor networks, and data analytics to track the state of a field's soil belonging to the household or a commercial crop to locate the individual positions of crops and their needs and monitor them. Administration of individual portions of fields can be compressed to centimeter-by-centimeter management of the exact state of a field, which generates limited stresses in crops and higher yields and can significantly minimize the costs incurred (Zain et al., 2024). Unfortunately, not all farms have adopted precision agricultural practices. The basic principle behind precision agriculture is the consideration that the crops grown are not the same and adjustments to a field management strategy are needed if one asset is to behave differently from another, and this can equate to lower costs and increased yields. Precision agriculture serves as a tool to help farmers best identify the requirements of their crops, separate the needs of differing crops grown on the same field, and assist mechanisms that ensure that the crop or field deficit coincides with the broader field (Jadhav et al., 2024; Liu et al., 2024). Each of these abilities, if optimally performed, can significantly reduce the amount of water or chemical waste produced by farming. These are summarized as the specific aims or benefits of adopting precision agricultural practices. The term "precision agriculture" became popular in the 1980s. It is an approach that has been made necessary by technological advances, driven primarily by the need for economic and environmental focus, including crop inputs (Rozenstein et al., 2024). The area of application of precision agriculture did not begin to clear until the global positioning system became available to enable the development of autonomous guidance systems. This development has now propelled precision agriculture to the forefront of popular concern. Precision agriculture not only becomes a part of modern farming, but its core functions can also help produce massive data that support new farming practices and real improvements in the outputs of farms (Gera & Jain, 2024; Pascutti et al., 2024).

1.4.2 Biotechnology

Biotechnology has been lauded as a potential game changer for agricultural production. A variety of methods that include genetic manipulations in the lab, like bioengineering, have been or are on the table. Genetically modified organisms (GMOs) are organisms whose genomes are modified by molecular techniques, usually to include a copy of a gene from another organism that gives the GMO an advantageous trait, as illustrated in Fig. 1.4. For instance, wild tobacco plants were recently used to grow strong and long carbon nanotubes via genetic modification (Li et al., 2024; Vishnoi & Goel, 2024). Advantages of GMO crops include improved crop yield, providing a means for increased and more sustainable production; pest resistance, which may become critical as climate change causes the range of some pests to expand; the ability to grow crops in harsher climates and environments, with a recent research focus on drought resistance; and biofortification of otherwise poor staple crops to prevent vitamin and mineral deficiencies in regions where such foods are primarily subsisted upon (Haydar et al., 2024; Obaid et al., 2024). Additionally, biotechnologies have enabled rapid crop development, much faster than traditional breeding standards.

There is a significant and persistent public reaction against the use of genetically modified organisms in agriculture, prompting stringent legislation in many parts of the world. In 2015, popular restrictions existed in 19 of 28 European Union countries, covering 65% of the population. Public and activist concerns over the use of such technology on an industrial scale are due to a number of possible ethical and environmental effects (Xun et al., 2024). The ecological impact concerns the possibility of gene escape, and there is a possible economic dependence upon commercial GMO distributors. On safety grounds, concerns exist over potential human health impacts in terms of challenging allergen profiles. The regulatory challenges in this domain are fundamentally based on the rapid technological developments that make legislation out of date by the time it is released, a strongly rooted public ethical framework against such use, and the precautionary principle under which decisions may be based when information is uncertain, and knowledge is evolving (Kayastha et al., 2024; Tokumaru et al., 2024). Nevertheless, despite these controversies, most of the key reviews in this domain agree that biotechnology may be used to go some way toward sustainable feeding of a world population estimated to reach greater than 9 billion by 2050 and in reducing the environmental burden on agricultural land use through a morally defensible approach (Jadhav et al., 2024).

1.5 Policy and Governance in Agriculture

Global and local policy structure the way agriculture is practiced. Policy directions and the means of adopting them reflect an aim to promote agricultural development and increase food security while protecting and enhancing the economic, social, and environmental well-being of future generations. Many of the issues that are considered under the heading of policy relate to governance more than technical policy (Liu et al., 2024). The guidelines state that policies pursued by governments to attain these varied goals inevitably also play a major role in shaping the future of agriculture.

The agreements were also part of a movement away from permitted interventions in trade, reducing or abolishing market price support. A farmer who receives a direct payment needs to comply with some conditions but may have more choices about what to produce. Most people aim to try to regain some of what has been lost in income through prices. The resolution aimed to reverse the decreasing trend of official development assistance for agriculture in many developing countries (Jadhav et al., 2024; Rozenstein et al., 2024). The so-called "Green Box" to support farmers who are in this position is restricted to just broadband policies and supply management mechanisms such as quotas to control surpluses. The important responsibilities for governance aspects include a focus on the establishment and functioning of three working groups. In many parts of the world, farmers' dependence on state support and subsidy is reducing. Some countries are usually cited as places where the government is starting to withdraw from supporting agriculture (Gera & Jain, 2024; Pascutti et al.,

2024). In many ways, getting the governance right in agriculture is one of the most intractable problems faced anywhere in the world.

1.5.1 International Agreements and Treaties

Agriculture has been shaped partly by international agreements and treaties, and this will likely continue in its commercial and trade practices and its contribution to consumption. The body of rules guiding international agriculture is dominated by treaties governing international trade. Although there are multilateral agreements that cover agriculture, it is also a concern for countries (Li et al., 2024; Vishnoi & Goel, 2024). Trade policies in agriculture are guided by a series of new verified multilateral and bilateral treaties, which address aspects of agriculture such as administration, food security, and trade.

Although there is no direct treaty that guides the internal organization and activity of agricultural production in different countries, the approval and acceptance of various international agreements and treaties have affected the agricultural practices of countries. These include environmental, health, and trade agreements. The guiding principles of international trade are flexibility, predictability, and fair practice (Haydar et al., 2024; Obaid et al., 2024). However, international trade treaties are not always flexible and predictable; they are not always summarized with fair practice for all. Not accepting agreements and conventions can limit the international development of many developing countries that depend on agriculture for consumption and their livelihoods. Compliance and enforcement of treaties are very important in international agreements (Tokumaru et al., 2024; Xun et al., 2024). This promotes equity in trade, as all members participate equally, thus supporting progress on the SDGs. National policies can no longer govern agricultural policies in many countries, as they affect several countries both directly and indirectly. Institutions of agriculture, including the Ministry and Department of Agriculture, are deeply affected by food and agriculture commitments in this country. Increasing agricultural production to feed the growing population can no longer be delayed. Countries must cooperate in matters of agriculture in order to sustain and maintain the consumption of food (Tokumaru et al., 2024). Negative factors affect the export of cheaper beef in Asian countries, while the production of beef in South American countries has also been negatively affected. Several developing countries have been impacted due to the demand for tuna in the international market. Trade in agriculture has become problematic, and several negotiations are taking place between countries regarding international trade. As more provisions cover agriculture, countries fail to agree on issues that are addressed in different treaties. Efforts have continued among international organizations to guide the development of trade in agriculture today (Kayastha et al., 2024; Mohammadi et al., 2024). This is why there is a need to cooperate as a group in dealing with food and agriculture and to include international trade in negotiations and discussions.

1.5.2 *Government Subsidies*

An extensive system of subsidies is designed to support the agriculture sector and farming in OECD countries. Subsidies come in a variety of forms, such as direct income payments, crop insurance, and input subsidies that may not provide income so much as the reduced cost to the farmer. The subsidies are often presented as a way to provide a social benefit to farmers who perform valuable and socially beneficial deeds, such as feeding the nation and providing rural lands. They are also a result of strong lobbying efforts from the farm lobby. These subsidies can support farm income, keep land for farming, and keep food prices low for the consumer (Dai et al., 2024; Nalluri & Karri, 2024). To the extent that extra support for socially desirable practices is beneficial, these supports can encourage sustainability on the farm.

From an economic standpoint, however, subsidies enhance farmer income, and this increase in income may stimulate expanded production or reduce the change in production practices. By making potential negative costs lower than they are to the farmers, this also results in a reduction in market prices. Finally, subsidies received in cash may imply the continuation of certain practices (Sharma et al., 2024). Subsidy support can create unfair competition between farmers who have not been promised these funds and those who have. Technical considerations of the reforms should also reflect on the capacity of the authorities to suit diverse farming environments and practices. This is because definitions of specific sustainable farming practices are context-specific, market prices might also be affected (Obaid et al., 2024; Xun et al., 2024).

It should be noted here that there are two ways in which subsidies can affect the market price. Starting first from the demand side, there may be extra demand for a product, which tends to boost existing market prices. On the other hand, from the supply side, there may be increased production that supplies a greater amount at each possible price and shifts the supply curve to the right. In general, because subsidies reduce the cost of production, the supply of a product increases, which will increase the market price (Tokumaru et al., 2024). Subsidies may be flat or adjusted by some relevant factor, such as the amount of land in production. Once again, more market distortion can occur over time by farmers who are expanding or shooting for subsidies by altering the factors that determine their eligibility. Finally, small farmers often have less access to administrative resources, which makes the application for subsidies particularly difficult (Jadhav et al., 2024; Shafik et al., 2023).

1.6 Future Directions in Food and Agriculture

The future trends in food and agriculture are

(1) Given the current and emerging challenges associated with intensifying agriculture-food systems and adapting to and mitigating global climate change, as well as the evolution of demand for diversification of agriculture, innovations are necessary for the development of agricultural systems. While some farmers will not be able to innovate or adapt, new avenues of farm support will be required to address some of the emerging challenges and to change the focus of farm governance (Jadhav et al., 2024).
(2) Imagine possibilities of megacities that are self-sufficient in vegetables, fruits, and fish for local consumption. Urban agriculture may be attracting the lead of support for the green roof, which is increasingly under discussion in some cities. The Environment Agency is looking at measures to assist in bringing rooftops and tall building developments into production locally (Liu et al., 2024).
(3) Another possibility would be vertical farming with hydroponic cultivation of plants and animals in multiple varieties of productive arrangements. There is a growing area with oil palm. Agroforestry as an agricultural system is gaining much attention as a sustainable alternative to full reliance on monoculture short-term farming (Rozenstein et al., 2024).
(4) The food movements spreading around the world include those who support local farmers, local foods, or organic foods and often center on sustainability. These movements are not short-term fashion trends but a result of long-term considerations. They cause change, and their positive impacts include the growth of certified organic and going beyond organic movements (Pascutti et al., 2024). Awareness is also key for some consumer food purchases—local fish being one example.
(5) In light of the gains from interdisciplinary project research, the focus of both research innovation and policy development in agriculture should be interdisciplinary. In reality, across several countries and agencies, disciplinary research has been the primary focus within research organizations, with subsequent projects being appended together (Gera & Jain, 2024).
(6) The global proportion of food production under irrigation continues to increase at an increasing rate. This fact makes questions about food security and prices not solely a matter of the socially differentiated distribution and allocation of food. The issue of water recedes in many contexts. However, a future trend might be that one area of future research may again place more emphasis on water quantity—will there be enough water to produce future food (Li et al., 2024)? The nature of this question might include water demand from other water uses.
(7) There may be a time when we will seriously address the issue of food resilience in the face of climate change. Research may then have to address the issue of food varieties and yield assessments that take into account not just mean yield production but variability in climate. Central to this is the idea that crop varieties may need to be developed to thrive in climates that do not feature as much stability (Vishnoi & Goel, 2024). When this issue is brought to the fore, we are yet entering another term of future research.

(8) Overall, potential future negotiations, agreements, and research around agriculture and food systems will have to reconcile agricultural needs with emerging concerns in society and business associations to develop a broad industry (Haydar et al., 2024). These are what are known as evolving societal needs, and thus, agriculture evolves to meet them as well.

1.7 Emerging Trends

A number of emerging trends are themselves modest, but their cumulative effects will significantly shape our future. Particularly in light of growing awareness of the need for radical change, they signal a direction: inclusive food systems and regenerative practices are becoming trendy among some urban communities, and it is essential to understand how this links to broader concerns with and orientations of the transition toward agroecology (Obaid et al., 2024; Xun et al., 2024). Urban Agriculture The rise in urban agriculture was originally fueled by an interest in local, low-impact food production in an era bookended by concerns about oil food security and climate change. As these are becoming more pressing concerns, local authorities and communities have been promoting local agriculture as part of their search for sustainability and food security. Vertical Farming Shifting the attention from the edges of cities to city centers, the next step is urban farming as a way of addressing food scarcity and land resources in cities (Kayastha et al., 2024; Tokumaru et al., 2024). Technological optimism imagines a vertical cityscape that is efficient in its use of water, energy, and other resources. Technology The new technology signposting the future is, in some ways, its solidification and isolation from commercially useful potentials. Autonomy and automation in farming are linked. Unmanned fields led early on to greater efficiency, and in the preceding decade, technology companies have positioned automated vertical farming as a ready-made solution to our current and future food problems (Mohammadi et al., 2024; Saikanth et al., 2024). Environmental Concerns Automated and infrastructure-rich imaginaries of the urban environment are increasingly synonymous with a dystopian and ecologically impoverished urban future. Against this backdrop, there are increasing trend bubbles that place more emphasis on regenerative practices in town as opposed to sealed technologies. Moreover, regenerative agroecological trends are as likely to be visible in urban milieus as in small towns, where productive permaculture and agroforestry are showing the use of land regeneration for local sustainability (Nalluri & Karri, 2024).

1.8 Research Priorities

Human well-being and environmental health are inseparably entwined; therefore, it is important to understand the interdisciplinary nature of complex agricultural systems. While this research focuses on agronomic solutions to improve the impact of agriculture on the environment, these solutions are socially, economically, and environmentally viable for the producer (Dai et al., 2024). We suggest a few areas for focused research moving forward. The first is resilience under climate and weather variability. Atmospheric CO_2 levels are higher, and violent weather fluctuations are becoming increasingly severe. A second area is in innovative solutions for producing food, fiber, and energy. Research in organic amendments, soil microbial life, low-impact pesticides, and fertilizers for both conventional and organic producers is needed (Jose et al., 2024; Sharma et al., 2024). Work in sustainable fruit and vegetable production and pest and disease prevention, either with or without genetic modifications, is needed. There are numerous direct human health and consumer issues, including antibiotic resistance in the environment. Furthermore, climate change and evolving pests will require the development of new storage plant varieties and cropping methods for cooler climates. Even with all of these methods, indoor horizontal agriculture may be needed to produce food for some underdeveloped countries (Yuniwati et al., 2024; Zain et al., 2024). Helping the sector learn from existing soil and waste management literature, as well as testing also needs to be done.

1.9 Conclusion

Research and extension in crop breeding for competitive alternatives in food, fiber, fuel, health, and energy. These varieties should also have consumer demand or savvy programming stating the benefits of these alternatives. Research and extension in continuously improving mechanical technologies to enhance the working lives of producers and laborers while also synchronizing with developing varieties and consumer demands is necessary. Research also needs to be developed in non-mechanical technologies for them. Scale-up research and development for all of these aspects, including research into the social implications of the above items and implementation of these alternatives using a living laboratory of both commercial and experimental techniques, is essential. This can also be referred to as the formal definition of citizen science or a public–private partnership, and it is well poised to involve companies as well as universities supplying scientific research. Furthermore, it can facilitate a return to the days of the historic land grant mission and the interdisciplinary cooperative model, as farmers and scientists work together in land grant extension. This has become a less frequent and more formal and commercial interaction. This is especially true as consumers find their economic and political power and voices once again disparaged, undervalued, or forgotten. Participants would likely want to know how they or their neighbors were also interacting with the project, and compiling and maintaining that information would also have important implications for social and equitable access to the possible impacts this research aims to produce.

References

Dai, Z., Wang, Q., Jiang, J., & Lu, Y. (2024). Influence of university agricultural technology extension on efficient and sustainable agriculture. *Scientific Reports, 14*(1). https://doi.org/10.1038/s41598-024-55641-1

Gera, R., & Jain, A. (2024). Harnessing IoT and machine learning for sustainable agriculture: Predictive crop yield modeling in smart farming. *Journal of Autonomous Intelligence*, *7*(4). https://doi.org/10.32629/jai.v7i4.1108

Haydar, M. S., Ghosh, D., & Roy, S. (2024). Slow and controlled release nanofertilizers as an efficient tool for sustainable agriculture: Recent understanding and concerns. *Plant Nano Biology, 7*. https://doi.org/10.1016/j.plana.2024.100058

Jadhav, I., Singh, S., & Joshi, A. K. R. (2024). Modifying Rhizobacteria for improved plant growth and soil health in sustainable agriculture. *Passer Journal of Basic and Applied Sciences, 6*(Issue 1). https://doi.org/10.24271/PSR.2023.399649.1334

Jose, S., Renuka, N., Ratha, S. K., Kumari, S., & Bux, F. (2024). Bioprospecting of microalgae from agricultural fields and developing consortia for sustainable agriculture. *Algal Research, 78*, 103428. https://doi.org/10.1016/j.algal.2024.103428

Kalinaki, K., Shafik, W., Gutu, T. J. L., & Malik, O. A. (2023). *Computer vision and machine learning for smart farming and agriculture practices*. https://doi.org/10.4018/978-1-6684-8516-3.ch005

Katti, J., Dharmale, G., Ubale, S. A., Deoghare, R., & Sutar, S. H. (2024). IoT innovations in cotton plant disease detection for sustainable agriculture. *International Journal of Intelligent Systems and Applications in Engineering*, *12*(15s).

Kayastha, S., Behera, A., Sahoo, J. P., & Mahapatra, M. (2024). Growing green: Sustainable agriculture meets precision farming: A review. *Bhartiya Krishi Anusandhan Patrika, Of*. https://doi.org/10.18805/bkap697

Li, X., Mao, N., Zhang, W., Wei, X., Liu, T., Cheng, J., & Shao, M. (2024). Appropriateness of introducing earthworms into sustainable agriculture from the perspective of soil carbon emissions. *Soil and Tillage Research, 237*, 105961. https://doi.org/10.1016/j.still.2023.105961

Liu, M., Chen, X., & Jiao, Y. (2024). Sustainable agriculture: Theories, methods, practices and policies. *Agriculture (Switzerland), 14*(Issue 3). https://doi.org/10.3390/agriculture14030473

Mohammadi, S., Jabbari, F., Cidonio, G., & Babaeipour, V. (2024). Revolutionizing agriculture: Harnessing nano-innovations for sustainable farming and environmental preservation. *Pesticide Biochemistry and Physiology, 198*. https://doi.org/10.1016/j.pestbp.2023.105722

Moreira, B. R. de A., Marra, T. M., Silva, E. A. da, Brito Filho, A. L. de, Barbosa Júnior, M. R., Santos, A. F. dos, Silva, R. P. da, & Vellidis, G. (2024). Advancements in peanut mechanization: Implications for sustainable agriculture. *Agricultural Systems, 215*. https://doi.org/10.1016/j.agsy.2024.103868

Nalluri, N., & Karri, V. R. (2024). Superior effect of nature based solutions in soil and water management for sustainable agriculture. *Plant Science Today, 11*(Issue 1). https://doi.org/10.14719/pst.2159

Nawaz, T., Saud, S., Gu, L., Khan, I., Fahad, S., & Zhou, R. (2024). Cyanobacteria: Harnessing the power of microorganisms for plant growth promotion, stress alleviation, and phytoremediation in the era of sustainable agriculture. *Plant Stress, 11*. https://doi.org/10.1016/j.stress.2024.100399

Obaid, M. K., Alazzai, W. K., Abood, B. S. Z., & Al-Farouni, M. (2024). Sustainable agriculture practices: AI and IoT's vital contribution. In *E3S web of conferences* (Vol. 491). https://doi.org/10.1051/e3sconf/202449101025

Pascutti, M. C. D., Silvestre, R. da S. F., & Ortiz, T. A. (2024). The role of microorganisms in sustainable agriculture: A review. *DELOS: DESARROLLO LOCAL SOSTENIBLE, 17*(52). https://doi.org/10.55905/rdelosv17.n52-001

Pszczółkowski, P., & Sawicka, B. (2024). Ultrasound application in potato cultivation: Potential for enhanced yield and sustainable agriculture. *Sustainability (Switzerland), 16*(1). https://doi.org/10.3390/su16010108

Rozenstein, O., Cohen, Y., Alchanatis, V., Behrendt, K., Bonfil, D. J., Eshel, G., Harari, A., Harris, W. E., Klapp, I., Laor, Y., Linker, R., Paz-Kagan, T., Peets, S., Rutter, S. M., Salzer, Y., & Lowenberg-DeBoer, J. (2024). Data-driven agriculture and sustainable farming: Friends or foes? *Precision Agriculture, 25*(1). https://doi.org/10.1007/s11119-023-10061-5

Saikanth, D. R. K., Ragini, M., Tripathi, G., Rajesh, K., Giri, A., Pandey, S. K., & Verma, L. K. (2024). The impact of emerging technologies on sustainable agriculture and rural development. *International Journal of Environment and Climate Change, 14*(1). https://doi.org/10.9734/ijecc/2024/v14i13830

Shafik, W. (2024a). Barriers to implementing computational intelligence-based agriculture system. In *Computational intelligence in internet of agricultural things* (pp. 193–219). Springer. https://doi.org/10.1007/978-3-031-67450-1_8

Shafik, W. (2024b). Biosensor-based drones anomaly detection integration for sustainable agriculture development. In *Achieving food security through sustainable agriculture* (pp. 283–316). IGI Global. https://doi.org/10.4018/979-8-3693-4240-4.ch012

Shafik, W., & Kalinaki, K. (2024). Societal and ethical implications of technology-enhanced agriculture and healthcare: An African context. In *2024 IST-Africa conference (IST-Africa)* (pp. 1–11). https://doi.org/10.23919/IST-Africa63983.2024.10569306

Shafik, W., Tufail, A., Apong, R. A. A. H. M., & De Silva, L. C. (2024a). Internet of Things for smart agricultural practices. In *Internet of Things applications and technology* (pp. 190–217). Auerbach Publications. https://doi.org/10.1201/9781003458401-13

Shafik, W., Tufail, A., De Silva Liyanage, C., & Apong, R. A. A. H. M. (2024b). Using transfer learning-based plant disease classification and detection for sustainable agriculture. *BMC Plant Biology, 24*(1), 136. https://doi.org/10.1186/s12870-024-04825-y

Shafik, W., Tufail, A., Liyanage, C. D. S., & Apong, R. A. A. H. M. (2023). Using a novel convolutional neural network for plant pests detection and disease classification. *Journal of the Science of Food and Agriculture*. https://doi.org/10.1002/jsfa.12700

Sharma, P., Thakur, N., Mann, N. A., & Umar, A. (2024). Melatonin as plant growth regulator in sustainable agriculture. *Scientia Horticulturae, 323*, 112421. https://doi.org/10.1016/j.scienta.2023.112421

Tokumaru, S., Tokushima, Y., Ito, S., Yamaguchi, T., & Shimoda, M. (2024). Advanced methods for insect nets: Red-colored nets contribute to sustainable agriculture. *Scientific Reports, 14*(1). https://doi.org/10.1038/s41598-024-52108-1

Vishnoi, S., & Goel, R. K. (2024). Climate smart agriculture for sustainable productivity and healthy landscapes. *Environmental Science and Policy, 151*, 103600. https://doi.org/10.1016/j.envsci.2023.103600

Xun, W., Liu, Y., Ma, A., Yan, H., Miao, Y., Shao, J., Zhang, N., Xu, Z., Shen, Q., & Zhang, R. (2024). Dissection of rhizosphere microbiome and exploiting strategies for sustainable agriculture. *New Phytologist*. https://doi.org/10.1111/nph.19697

Yuniwati, E. D., Hartati, F. K., Sutanto, A., Sahabuddin, Kustini, Siswati, N. D., & Iskandar, T. (2024). Biochar technology to increase cassava crop productivity: A study of sustainable agriculture on degraded land. *Open Agriculture, 9*(1). https://doi.org/10.1515/opag-2022-0260

Zain, M., Ma, H., Ur Rahman, S., Nuruzzaman, M., Chaudhary, S., Azeem, I., Mehmood, F., Duan, A., & Sun, C. (2024). Nanotechnology in precision agriculture: Advancing towards sustainable crop production. *Plant Physiology and Biochemistry, 206*. https://doi.org/10.1016/j.plaphy.2023.108244

Chapter 2
Food and Water Sustainability: Challenges and Prospects

Md. Mehedi Hasan Sawon, Mohammad Badruddoza Talukder, and Md. Akram Hossain

Abstract Climate change, resource shortages, and environmental degradation have caused a great global food and water crisis in recent eras. Population growth, urbanization, and increasing freshwater consumption also threaten food security and ecological health. Hence, understanding the contribution of food and water sustainability is crucial for sustainable development strategies and resource allocation. This chapter aims to find out the synergies and trade-offs between food and water sustainability and environmental consequences to create sustainable development and sustainable resource allocation strategies by evaluating the contributions of water and food sustainability to environmental protection. The study highlights the interconnectedness of food, water, and the Environment, highlighting significant challenges of population growth, urbanization, climate change, and ecological degradationlenges. It suggests solutions, including technical advances, legislative changes, and community-based methods to overcome food and water sustainability challenges. It emphasizes the importance of integrated water resource management, agroforestry, organic farming, and climate change adaptation to ensure food and water sustainability, which will contribute to achieving sustainable development goals in the long run.

Keywords Sustainability · Food security · Water sustainability · Sustainable development goals · Systemic food-water-environment interrelationships · Resource optimization · Justice and equity · Policy reforms · Sustainable production

Md. M. H. Sawon (✉)
Department of Tourism and Hospitality Management, Daffodil International University, Dhaka, Bangladesh
e-mail: mehedisawon60@gmail.com

M. B. Talukder
College of Tourism and Hospitality Management, International University of Business Agriculture and Technology, Dhaka, Bangladesh

Md. A. Hossain
Department of Public Administration, University of Dhaka, Dhaka, Bangladesh

J. Parray and A. K. Haghi (eds.), *Food and Water Security*,
https://doi.org/10.1007/978-3-032-04870-7_2

2.1 Introduction

In recent years, climate change, resource shortages, and environmental degradation have stressed an integrated strategy to manage water and food systems worldwide (Vanham & Leip, 2020). The interconnectivity of various elements like water, energy, food, and the Environment emphasize the need for sustainable and fair solutions to the food and water crisis (Shah et al., 2024; Zahedi et al., 2024). Population growth, urbanization, and climate change are also worsening the global crisis related to food and water (Gupta et al., 2021). A holistic and interrelated strategy for sustainable development is needed due to food, water, and energy shortages and competition caused by climate change, urbanization, and economic pressures (Salem et al., 2022). The global food and water crisis, caused by population growth and food habit changes, has increased demand for edible food, fresh water, and cultivable land (Hemathilake & Gunathilake, 2022). According to Zahedi et al. (2024), freshwater consumption will triple by 2025 due to agricultural and industrial development. This growing shortage also threatens food and water security, economic growth, and ecological health (Bradu et al., 2023).

However, the global food and water landscape is rapidly changing toward cleaner, more sustainable sources (Zohuri, 2023). This paradigm shift provides prospects but presents problems concerning food and water availability and allocation. The food and water sustainability concerns also show how multiple elements of the Environment affect human well-being and social development (Henderson & Loreau, 2023). Agriculture, production, and industry need huge portions of water, whereas water treatment, pumping, and distribution require significant energy. Hence, all sectors, like energy and water, depend on ecosystem services from the Environment to sustain human activity and biodiversity (Karamian et al., 2021). Effective and sustainable solutions to food and water crises require understanding food and water sustainability to ensure human well-being and longevity (Henderson & Loreau, 2023).

This chapter examines the synergies and trade-offs between food and water sustainability and environmental consequences to create sustainable development and fair resource allocation strategies by evaluating water, food, and environmental interdependencies. This study covers the key challenges and ways forward to ensure food and water sustainability and sustainable development goals. This chapter also explores the issues of managing water and food needs and how to overcome the challenges to achieve Sustainable development goals.

2.1.1 Conceptual Framework

Complex connections and interdependence exist between food, water, and the Environment and their contributions to the SDGs (Scharlemann et al., 2020). Agricultural practices affect water quality and availability, which is needed for irrigation. Food production and water supply depend on environmental resources and ecological services (Rocha et al., 2020).

Several elements contribute to the food-water-environment sustainability, leading to sustainable development (Fig. 2.1), such as

(a) ***Systemic Interrelation***: Food, water, and the Environment are interrelated in this view, considering these aspects' interconnections and feedback loops (Kretschmer & Kahl, 2021).
(b) ***Resource Efficiency***: Optimizing food, water, and environmental resources to reduce waste and boost output by developing and implementing innovative resource-efficient technologies and techniques (Skawińska & Zalewski, 2022).
(c) ***Equity and Justice***: There is a need for equal food, water, and environmental access for disadvantaged groups by ensuring social and economic sustainability to share resource management advantages equally (Menton et al., 2020).
(d) ***Climate Change Resilience***: Climate change effects on food, water, and the Environment should be mitigated by developing climate change adaptation and food and water resilience methods (Srivastav et al., 2021).

Challenges to food-water-environment sustainability include

(a) ***Water Scarcity***: Population growth, urbanization, and climate change are worsening water shortages in many countries, causing difficulties for food production and other industries (Zhang et al., 2020).
(b) ***Food Insecurity***: Millions of people worldwide are malnourished and hungry due to economic reasons, climate change, and water shortages (Goncharova & Merzlyakova, 2021).
(c) ***Environmental Degradation***: Deforestation, pollution, and unsustainable agriculture are destroying ecosystems and threatening food and water security (Kumar et al., 2022).

Despite these obstacles, sustainable growth has the following prospects:

(a) ***Technological Advancement***: Advancements in technology in agriculture, water management, and renewable energy can improve resource efficiency and reduce environmental impacts (Khan et al., 2021).
(b) ***Policy Reforms***: Reforms of Policies can promote sustainable food and water management, ensuring equitable access and addressing environmental concerns (Nhamo et al., 2022).
(c) ***Community-Based Approaches***: Community-led efforts can help local communities manage resources sustainably and increase food security (Herrmann et al., 2020).

Several ways can solve food-water-environment nexus concerns and grasp the prospects:

(a) ***Integrated Water Resource Management***: This method considers agriculture, ecology, and other sectors when managing water supplies (Jain & Singh, 2023).

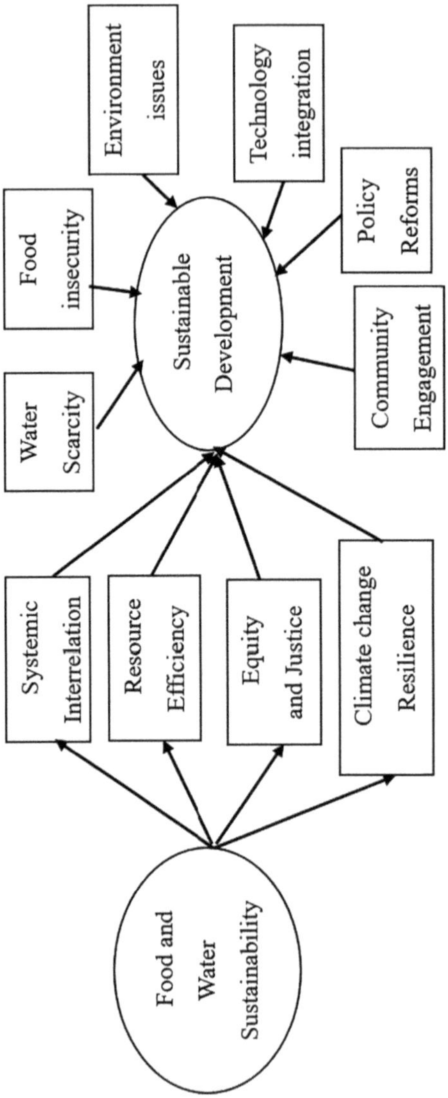

Fig. 2.1 Conceptual framework of food and water sustainability

(b) ***Sustainable Agriculture***: Agroforestry, organic farming, and conservation agriculture can boost food production and conserve the Environment (Saikanth et al., 2023).
(c) ***Climate Change Adaptation***: Creating and executing climate change adaptation strategies can reduce food and water system-related negative consequences (Gomez-Zavaglia et al., 2020).
(d) ***Policy Coherence***: Effective implementation and lasting results need policy coherence in agriculture, water, Environment, and climate change (Beretić et al., 2024).

2.2 Food and Water Sustainability Contribution to Sustainable Development Goals

2.2.1 Systemic Interrelation Between Food, Water, and Environment

Food, water, and the Environment are interconnected, forming the basis of sustainable development (Scharlemann et al., 2020). Agriculture is essential to food production, i.e., crop irrigation uses a lot of water from rivers, lakes, and groundwater (Venkatesharaju, 2022). Population growth, changing diets, and development in irrigation increase agricultural water demand (De Wrachien et al., 2021). Food production and water usage show the delicate balance needed for sustainable resource usage (Li et al., 2022a, 2022b). Agricultural operations significantly affect water quality as pesticides, fertilizers, and other agrochemicals can pollute water, endangering humans and ecosystems (Zahoor & Mushtaq, 2023). Deforestation and soil erosion cause sedimentation and nutrient runoff, worsening water quality. Hence, the environmental impacts of agriculture need careful attention and sustainable cultivation (Lal, 2020).

Food production has environmental consequences beyond water quality. For instance, agriculture-related deforestation causes climate change and biodiversity loss (Movilla-Pateiro et al., 2021). Food processing and transportation utilize energy and emit greenhouse gases. These environmental costs of food production must be balanced against food security and economic growth (Rehman et al., 2022). Complex feedback loops exist between food, water, and the Environment since water shortage reduces food production, raising costs and resource demand (Zeng et al., 2022). A vicious cycle of water shortages and environmental deterioration often results from unsustainable farming methods. Practical solutions to control these aspects' interdependence require understanding these feedback loops (Zhang et al., 2020).

Food, water, and the Environment are interconnected, affecting sustainable development (Wang et al., 2021). Hence, food security, water availability, and environmental preservation must be addressed holistically (Mishra et al., 2021). Sustainable development objectives for poverty eradication, hunger, clean water and sanitation, and climate action are interconnected with food, water, and environmental resource management (Rai, 2023). A systems-based approach to sustainable development is needed since food, water, and the Environment are interconnected (D'Amore et al., 2022). This requires careful resource allocation, environmental preservation, sustainable agriculture, and other activities (Muhie, 2022).

2.2.2 *Resource Efficiency for Food, Water, and Environmental Optimization*

Sustainable development relies on resource efficiency to decrease waste and optimize production from food, water, and natural resources (Yue et al., 2021). Resource efficiency reduces resource consumption and boosts production, promoting environmental sustainability, economic development, and social well-being (Mikhno et al., 2021). Precision agricultural methods like satellite imaging and GPS-guided irrigation systems help farmers save water and fertilizer and increase yields (Gawande et al., 2023). Recent food processing and packaging advances reduce food waste and shelf life, lowering food production and delivery's economic and environmental impact (Brennan et al., 2021; Talukder et al., 2024a, 2024b, 2024c). Resource efficiency is greatly improved by sustainable agriculture. For instance, crop rotation, cover cropping, and agroforestry promote soil-healthy food production and water retention (Fahad et al., 2022). Integrated pest control reduces pesticide usage, conserving water quality and the Environment (Paudel et al., 2020). Diversifying agricultural systems promotes ecological resilience and reduces crop failures, improving resource efficiency (Dardonville et al., 2022).

Innovative water conservation devices can cut water use across industries. For instance, drip irrigation hydrates plant roots directly, reducing evaporation and runoff (Abou-Shady et al., 2023). Wastewater treatment and recycling reuse water in tourism, agriculture, energy and other industries (Talukder et al., 2024a, 2024b, 2024c). Water waste can be reduced through efficient water metering and leak detection (Varma, 2022). By reducing waste and increasing resource value, circular economy concepts improve resource efficiency (Tambovceva et al., 2021). Designing items and systems for reuse, recycling, and repair reduces resource extraction. By conserving resources, circular economy techniques enhance food production, packaging, and waste management (Gonçalves & Maximo, 2023; Kumar et al., 2024).

Avoiding food waste through storage, planning, and consumption saves resources. Consumers reduce food waste and choose local and seasonal goods through eco-friendly education and awareness efforts (Bolwig et al., 2021). Resource efficiency requires good policies and governance, and Governments should invest in research and development, encourage resource-efficient technology, and regulate waste and pollution (Luo et al., 2024). Sustainable development relies on resource efficiency to cut waste, save resources, and boost production (Pimenov et al., 2022). Thus, innovative technology, sustainable practices, and effective regulations optimize food, water, and environmental resource utilization, guaranteeing a more sustainable and resilient future for future generations (McClements et al., 2021).

2.2.3 Justice and Equity in Food, Water, and Environmental Access

Sustainable development relies on equity and justice, as equitable access to food, water, and the Environment is critical for tackling social and economic imbalances (Berry, 2020). Indigenous peoples, disadvantaged communities, and low-income populations suffer significant problems obtaining food, water, and natural resources (Rajapakse et al., 2022). Due to inadequate economic resources, landlessness, and discriminatory legislation, these communities are more exposed to food poverty, water shortages, and environmental damage (Birkmann et al., 2022).

Resource insecurity harms underprivileged populations economically and socially by causing malnutrition, health issues, and lower education (Zelber-Sagi et al., 2024). Water shortage can reduce access to drinking, sanitation, and hygiene, hurting health and productivity (Zakar et al., 2020). Food, water, and natural resources must be distributed fairly to provide equal access by evaluating the needs of diverse groups, particularly the most vulnerable, and allocating resources somewhat based on population, land ownership, and previous disenfranchisement (Logan & Guikema, 2020; Talukder et al., 2024a, 2024b, 2024c).

Equity and justice in resource management require empowering underrepresented populations. This includes providing education, training, financial resources, and support for community-based organizations that serve underprivileged people (Marrone et al., 2022). Empowering disadvantaged groups helps them make decisions and push for resource access (Coy et al., 2021). Resource management must address all groups' demands through inclusive governance by creating participatory procedures for underprivileged populations to help design and execute policies (Roka, 2020). Policymakers can make more fair and effective resource management policies by including disadvantaged populations (Fernandez-Bou et al., 2021).

Past drawbacks must be addressed with specific policies and initiatives to help disadvantaged people engage more fully in society (Salmi & D'Addio, 2021). Land reform, restitution, and infrastructure and service improvements in historically disadvantaged communities are needed (Delius & Schirmer, 2021). Sustainability in the food-water-environment nexus requires equity and justice (Caron, 2023). A more fair and sustainable society can be achieved by addressing the needs of disadvantaged populations and fostering equitable resource access, necessitating inclusive government, marginalized community empowerment, and historical disadvantage reparation (Amri & Sihotang, 2023).

2.2.4 Climate Change Resilience for Adapting to Changing Landscape

Climate change threatens food, water, and the Environment worldwide (Muluneh, 2021). Rising temperatures have changed precipitation patterns, and severe weather events alter ecosystems, reduce agricultural production, and worsen water shortages (Nhemachena et al., 2020). Sustainable development and future generations need climate change resistance (Poli et al., 2024). Environmental changes caused by climate change influence food, water, and ecosystems (Prakash, 2021). Temperature increases can affect crop yields, insect and disease outbreaks, and agricultural production zones (Skendžić et al., 2021). Droughts, floods, and water availability can impair agricultural productivity and clean water supply due to altered precipitation patterns (Nhemachena et al., 2020). Hurricanes, typhoons, and heatwaves devastate infrastructure, agriculture, and ecosystems (Loucks, 2021).

Climate change threatens agricultural production as rising temperatures impact crop growth, development, and yield (Malhi et al., 2021). Changes in precipitation patterns can affect agricultural irrigation and soil moisture. Dry spells and floods can hinder crop productivity and cause crop losses (Grusson et al., 2021). Climate change affects water supply and quality. Some locations can experience water shortages and floods due to precipitation changes (Amanullah et al., 2020). River and lake flows can be affected by glacier melt and snowpack loss due to rising temperatures (Du et al., 2022). Droughts and floods can become more frequent and intense due to climate change, further affecting water supplies (Tabari, 2020).

Climate change disrupts and weakens ecosystems. For instance, high temperatures can change species ranges, disturb ecological processes, and raise extinction risk (Upadhyay, 2020). Wetland, coastal, and freshwater ecosystems can be affected by precipitation patterns. Extreme weather can harm ecosystems, reducing biodiversity and habitat (Hassan et al., 2020). Climate change adaptation must address food, water, and ecological effects to build resilience. Important tactics include:

(a) ***Agricultural Diversification***: Growing climate-resistant crops reduces crop failures and ensures food security (Singh et al., 2022).
(b) ***Improved Water Management***: Rainwater collection, drip irrigation, and recycling can preserve water and lessen agricultural systems' drought sensitivity (Ray & Majumder, 2024).
(c) ***Conservation Agriculture***: Sustainable farming practices that conserve soil health, prevent erosion, and promote water retention can boost climate change resistance (Cárceles Rodríguez et al., 2022).
(d) ***Ecosystem Restoration***: Restoring damaged ecosystems can increase climate change resistance and offer ecosystem services like water filtering and flood control (Li et al., 2022a, 2022b).

Developing early warning systems for severe weather events can help farmers and communities prepare for and prevent climate change (van Ginkel & Biradar, 2021). Climate change resilience requires good policy and governance. Addressing climate change and sharing information and best practices requires international collaboration (Owen, 2020).

2.3 Challenges to Ensure Food and Water Sustainability Contributions to SDG

2.3.1 Water Scarcity Affecting Food and Water Sustainability

Population increase, urbanization, and climate change worsen worldwide water shortage (Zhang et al., 2020). Water supply decreases as demand rises, causing issues for food production, industry, and other sectors (Li et al., 2022a, 2022b). The growing global population, especially in emerging nations, causes water shortages. Population increases water demand for home, agricultural, and industrial usage (Zahoor & Mushtaq, 2023). City life increases water demand since municipal services like drinking water, sanitation, and industrial activities need enormous amounts of water (Lal, 2020). Climate change is modifying precipitation patterns and worsening severe weather, causing water shortages and abundant issues (Movilla-Pateiro et al., 2021). Some locations experience droughts, while others flood due to temperature and rainfall changes. Rising temperatures can pollute water and lower water quality due to climate change (Rehman et al., 2022).

The greatest water user worldwide is agriculture, at 70%. Demand for agricultural water rises with food demand. This can cause groundwater over-extraction, depletion, and allocation disputes (Zhang et al., 2020). Manufacturing, energy generation, and mining need plenty of water. High water demand in these industries leads to rivalry with other users in places with limited water supplies (Zhang et al., 2020). Water shortage harms the ecosystem. For instance, land subsidence and saltwater intrusion from groundwater depletion can damage ecosystems and agriculture (Wang et al., 2021). Water scarcity makes aquatic habitats more susceptible to stress and deterioration, affecting biodiversity (Mishra et al., 2021).

Water shortage has serious economic and social effects as nations depend on agriculture, yet water shortages can diminish agricultural output, food insecurity, and rural poverty (Rai, 2023). Water shortage can also slow industrialization and economic progress. Clean water is needed for sanitation, hygiene, and disease prevention; hence water shortage harms public health (D'Amore et al., 2022). Population increase, urbanization, and climate change worsen worldwide water shortage, demanding a holistic strategy encompassing water, food, and the Environment. For a sustainable future, water management must include conservation, efficiency, and fair distribution (Muhie, 2022).

2.3.2 Food Insecurity Affecting Food and Water Sustainability

Millions worldwide struggle with food insecurity and the lack of enough safe and nutritious food to fulfill basic requirements (Yue et al., 2021). Economic factors contribute to food insecurity. For instance, poverty, economic inequality, and unemployment can restrict food availability for vulnerable groups (Mikhno et al., 2021). Production, transportation, and market speculation increase food prices, making it unaffordable for many (Gawande et al., 2023). Food insecurity increases during economic downturns and crises due to lower family incomes and food availability (Brennan et al., 2021). Agriculture is affected by climate change, impacting yields, quality, and availability. For instance, the growing temperatures can affect crop growth, development, and yield (Malhi et al., 2021). Rainfall patterns can cause droughts, floods, and soil erosion, interrupting agriculture and lowering food harvests. Hurricanes, typhoons, and heatwaves destroy crops and agricultural infrastructure (Du et al., 2022).

Water limitations hinder food production. Agriculture uses the most water; as water grows scarcer, meeting food demand becomes harder (Abou-Shady et al., 2023). Limited irrigation, agricultural yields, and production costs result from water shortages. Water shortages compel farmers to quit their land, degrading it and worsening food poverty (Varma, 2022). Conflicts and relocation can devastate food security. Conflicts can damage infrastructure, food production, and markets (Tambovceva et al., 2021). Food shortages and starvation can result from displacement. Food and other necessities are needed by displaced persons in numerous crisis zones (Gonçalves & Maximo, 2023).

Social and economic hardships result from food insecurity. Malnutrition can also cause stunted development, cognitive impairment, and illness vulnerability (Zelber-Sagi et al., 2024). Hunger can also hinder education and production, restricting people and communities. Food insecurity can result from social unrest, political instability, and poverty (Fernandez-Bou et al., 2021). Food insecurity is diverse and needs a holistic response. Poverty, climate change, and water shortages must be addressed to ensure food security for everyone. Economic, social, and environmental measures and worldwide solidarity are needed (Amri & Sihotang, 2023).

2.3.3 Environmental Degradation Threatening Food and Water Sustainability

Deforestation, pollution, and unsustainable agriculture threaten food and water sustainability (Skendžić et al., 2021). Ecosystem loss and natural resource deterioration reduce the Environment's ability to deliver essential services for human well-being (Nhemachena et al., 2020). Clearing forests is a significant cause of environmental decline. Forests regulate temperature, maintain biodiversity, filter water, and limit erosion. Soil erosion, river sedimentation, and biodiversity loss can result from deforestation. It releases stored carbon dioxide into the atmosphere, contributing to climate change (Loucks, 2021).

Another major environmental issue is pollution, which contaminates air, water, and soil. Industrial, agricultural, and household waste can pollute water sources, rendering them hazardous for drinking and irrigation (Muluneh, 2021). Air pollution harms health and causes climate change. Pollution reduces soil fertility and agricultural output (Grusson et al., 2021).

Pesticide, fertilizer, and irrigation overuse can degrade the Environment. Water, animal, and soil degradation can result from pesticide usage (Hassan et al., 2020). Deforestation and overgrazing cause desertification and soil erosion. Groundwater depletion and soil salinization can result from over-irrigation (Singh et al., 2022).

Environmental deterioration threatens food and water security: soil erosion and lower agricultural productivity from deforestation harm food production (Ray & Majumder, 2024). Pollution contaminates water sources, rendering them hazardous for drinking and irrigation and harming fishing ecosystems. Unsustainable agriculture can erode soil, diminish agricultural yields, and deplete water (Cárceles Rodríguez et al., 2022).

Ecosystems offer many vital services to humans. These include water filtration, climate control, pollination, and biodiversity protection (Li et al., 2022a, 2022b). Environmental degradation can harm ecosystem services and food and water security. As trees control climate and filter water, deforestation and pollution can damage aquatic ecosystems that provide fisheries and clean water.

Deforestation, pollution, and unsustainable agriculture threaten food and water sustainability (van Ginkel & Biradar, 2021). Ecosystem loss and resource degradation reduce the Environment's ability to deliver essential services for human well-being. Sustainable land management, pollution control, and sustainable agriculture are needed to combat environmental deterioration (Owen, 2020).

2.4 Prospects for Food and Water Sustainability Contributions to SDG

2.4.1 Technological Advancement Promoting Sustainable Food and Water Sustainability

Technology is essential for sustainable food and water management. Agriculture, water management, and renewable energy innovations can improve resource efficiency, decrease environmental consequences, and support the SDGs (Khan et al., 2021), such as

(a) ***Precision Agriculture***: Satellite imaging, GPS, and sensors help farmers maximize resource usage, reduce waste, and boost yields. Precision agriculture improves crop quality and reduces pesticide, fertilizer, and water consumption (Pandey et al., 2021).
(b) ***Biotechnology***: Biotechnology can create pest-, disease-, and climate-resistant crop types, minimizing chemical inputs. Genetically modified crops can boost productivity and nutrition (Ameen et al., 2023).
(c) ***Vertical farming***: Indoor farming techniques save land and water usage by growing crops in controlled surroundings. Vertical farming allows year-round production, cutting food miles and providing fresh crops (Avgoustaki & Xydis, 2020).

Seawater can be converted into freshwater using desalination technology, providing a stable water supply in areas with little freshwater (Ashraf et al., 2021). Desalination is becoming more efficient and cost-effective, making it a feasible choice for many nations. Improved water treatment technology can cleanse and reuse wastewater, decreasing freshwater usage and pollution (Herrero et al., 2021). Wastewater treatment can create renewable energy via anaerobic digestion. The harvesting and groundwater recharge systems can gather and store rainfall, lowering surface water use and enhancing water security (Lajoie-O'Malley et al., 2020).

Solar electricity can power farms, water pumps, and food processors. Solar electricity can decrease fossil fuel consumption and greenhouse gas emissions (Pawlak & Kołodziejczak, 2020). Wind power is a clean and sustainable source of electricity for agricultural and water management. Regions with intense and persistent winds benefit from wind power. Bioenergy from biomass, such as agricultural waste, can heat, cook, and generate power. Bioenergy reduces fossil fuel use and promotes circular economy (Sherwood, 2020). Thus, innovative technology can help promote sustainable food and water management. Research, development, and deployment of breakthrough technologies can improve resource efficiency, minimize environmental consequences, and help achieve the SDGs (Qaim, 2020).

2.5 Policy Reforms Promoting Food and Water Sustainability

Effective policy changes are needed for sustainable food and water management, fair resource access, and environmental protection (Galli et al., 2020). Policy changes can help achieve the Sustainable Development Goals (SDGs) and a more sustainable future. Such as:

Policies on Agriculture

Reforming agricultural subsidies can encourage sustainable farming and lessen environmental consequences by phasing out payments for toxic pesticides and poor land management and raising subsidies for sustainable agriculture (Singh & Nath, 2020).

Land Reform: Land reform can help disadvantaged groups engage in agriculture by addressing land ownership and access inequities. Land reform can minimize deforestation and improve sustainable land management (Nhamo et al., 2022).

Trade Policies: Trade policy can encourage sustainable food production and consumption. Trade agreements can promote sustainable agricultural goods and minimize tariffs and subsidies that promote unsustainable food production and consumption (Smith & Glauber, 2020).

Policies on Water

Water Price: Water price can minimize waste and promote water efficiency. Revenue from water pricing can be utilized to enhance water quality and infrastructure (Beecher, 2020).

Water Allocation Policies: Equitable water allocation policies can allocate water resources equally across agriculture, industry, and households. These initiatives can also reduce water shortages and disputes (Sohrabi et al., 2022).

Integration of Water Resources: Integrated water resource management (IWRM) incorporates water, land, and ecosystem interactions. IWRM promotes sustainable and equitable water usage (Benson et al., 2020).

Environmental Policies

Pollution Control: Strict pollution control techniques safeguard water quality and decrease industrial and agricultural environmental consequences. Examples include wastewater discharge, pollutants, and pesticide and fertilizer controls (Wato et al., 2020).

Biodiversity Conservation: Conservation of biodiversity helps preserve ecosystem services like water filtration and pollination, which are necessary for sustainable food and water management. Examples include protected areas, habitat restoration, and endangered species protection (Oguh et al., 2021).

Policies on Emissions: Reduce greenhouse gas emissions to prevent climate change's effects on food and water supplies. This includes switching to renewable energy, improving energy efficiency, and minimizing deforestation (Anser et al., 2020).

Global food and water sustainability requires international collaboration. International agreements and organizations can reduce transboundary water challenges, improve sustainability, and share knowledge and technology (Siegel & Lima, 2020). Thus, the SDGs and sustainable food and water management need effective policy changes. Policy improvements can make the future more sustainable and fairer by tackling food and water insecurity and fostering resource equity (Borsellino et al., 2020).

2.6 Community-Based Approaches Empowering Local Communities for Sustainability

Sustainable food and water management is promising with community-based techniques. These initiatives can improve resource efficiency, environmental impact, and food security by allowing local people to control their resources and make decisions (Muhamad Khair et al., 2020). Community irrigation systems, water user groups, and natural resource management programs are examples of community-led efforts. These programs leverage local knowledge and expertise to design, implement, and manage resource utilization for sustainable solutions (Agustang et al., 2021). Local communities have resource management expertise. Traditional traditions and techniques frequently fit their environmental and social contexts. Community-based resource management systems can be sustainable and culturally suitable by utilizing local knowledge and creativity (Chatkaewnapanon & Lee, 2022).

Community-based initiatives provide local communities with a say in decisions. Communities feel greater ownership of their resources and embrace sustainable practices when they participate in planning and execution. Community members become more responsible and accountable for the long-term sustainability of their resources via empowerment. Community-based resource management can enhance social inclusion and equality by including all community members. This can correct past injustices and ensure disadvantaged groups get resources and advantages. Community-based initiatives can make societies fairer and more sustainable by increasing social cohesiveness and equality (Dolezal & Novelli, 2022). Community-based initiatives can strengthen climate change and other crisis resilience. Communities can resist external threats by diversifying livelihoods, fostering sustainable agriculture, and protecting natural resources. Resilience requires strong social networks and support systems, which community-based projects can provide (Wikantiyoso et al., 2021).

Community-based methods for sustainable resource management have promise but also drawbacks. Resource constraints, capacity issues, and community tensions can be issues. Community-based projects can promote sustainable growth by overcoming these obstacles and providing proper support (Mtapuri et al., 2022). For sustainable food and water management, community-based alternatives are promising. These initiatives can improve resource efficiency, environmental impact, and food security by empowering local people, using local knowledge, and fostering social inclusion. Governments and international organizations can promote sustainability and equity by supporting community-based projects (Kohsaka & Rogel, 2021).

2.7 Ways Forward to Solve Food and Water Sustainability Challenges

2.7.1 Integrated Water Resource Management (IWRM)

Integrated Water Resource Management (IWRM) integrates water, land, and ecosystem interactions (Nagata et al., 2022). IWRM uses a comprehensive approach to allocate water resources sustainably and fairly, balancing industry requirements and environmental protection, such as

IWRM Follows Various Principles

Participation: IWRM stresses stakeholder involvement in decision-making. IWRM involves government agencies, communities, corporations, and environmental groups to guarantee different viewpoints and socially acceptable choices (Sigalla et al., 2021).

Equity: IWRM ensures that vulnerable and disadvantaged communities receive a fair share of water resources. This requires correcting past injustices and allocating water resources fairly (Meran et al., 2021).

Efficiency: IWRM optimizes water usage to reduce waste and maximize benefits. Use efficient technology, reduce water losses, and promote water conservation (Kumar & Thakur, 2024).

Sustainability: IWRM protects the Environment and ensures sustainable water usage for future generations. This requires evaluating long-term water management implications and protecting ecosystems and biodiversity (Makanda et al., 2022).

IWRM Has Numerous Significant Parts

IWRM involves a thorough water resource evaluation, including availability, quality, and demand (Ngene et al., 2021). This data helps plan water management and distribute resources.

Water Allocation: IWRM allocates water resources to agriculture, industry, residential usage, and ecosystems. This necessitates assessing conflicting water needs and allocating water fairly and sustainably (Meran et al., 2021).

Water Conservation and Efficiency: IWRM encourages water conservation and efficiency to decrease water use and waste. This involves adequate irrigation, water leak reduction, and conservation (Upadhyaya, 2020).

Water Quality Management: IWRM handles pollution and contamination. Protecting water sources, treating wastewater, and monitoring water quality are required (Koop et al., 2022).

Ecosystem Protection: Protection of the Ecosystem Wetlands, forests, and groundwater recharge regions offer water-related benefits, and IWRM protects them. This requires ecological conservation and restoration (Kalogiannidis et al., 2023).

IWRM's Impact on Water Sustainability

Improved Water Security: IWRM can increase water security and alleviate water shortages by allocating water resources sustainably and fairly (Sigalla et al., 2021).

Enhanced Agricultural Production: IWRM can boost agrarian output by ensuring farmers have enough water and encouraging optimal water usage (Meran et al., 2021).

Reduced Environmental Degradation: By ensuring water resources are managed responsibly, IWRM can safeguard ecosystems and decrease environmental deterioration (Makanda et al., 2022).

Improved Public Health and Economic Growth: IWRM can promote public health by providing access to clean water and sanitation and can help economic growth by maintaining a stable water supply for industry and companies (Kumar & Thakur, 2024).

Therefore, comprehensive integrated water resource management helps provide sustainable and equitable water availability. IWRM promotes water efficiency, environmental protection, and sustainable development by considering water, land, and ecosystems (Koop et al., 2022).

2.7.2 Sustainable Agriculture for Food and Water Sustainability

Sustainable agriculture combines productivity, environmental care, and social equality. Sustainable techniques help farmers increase food output, save resources, and strengthen agricultural systems (Li et al., 2022a, 2022b).

Agroforestry

Agroforestry combines trees with crops or animals. Trees can improve soil health, biodiversity, climate change mitigation, and water management by improving soil structure, reducing erosion, and increasing soil fertility. Agroforestry systems can support a greater diversity of plants and animals, enhancing ecosystem resilience (Movilla-Pateiro et al., 2021).

Organic Farming

Organic farming practices, like crop rotation and cover cropping, can improve soil health, reduce environmental impacts, and enhance food quality. It can also support rural economies by creating jobs and supporting rural economies (Zhang et al., 2020).

Conservatism in Agriculture

Conservation agriculture can enhance soil health by improving structure, reducing erosion, and increasing fertility. It can also save water by reducing runoff and increasing infiltration, resulting in more effective water usage. Conservation agriculture can reduce greenhouse gas emissions by improving soil carbon sequestration and increasing agricultural productivity by improving soil health and reducing crop losses (Mishra et al., 2021).

Therefore, agroforestry, organic farming, and conservation agriculture can help solve food and water sustainability issues. These approaches boost agricultural output, preserve natural resources, and strengthen agrarian systems. Agriculture must be sustainable to provide food security, environmental protection, and rural development (Muhie, 2022).

2.7.3 Climate Change Adaptation for Food and Water System Resilience

Climate change threatens food and water systems worldwide. Effective climate change adaptation techniques are needed to combat climate change and ensure food and water security. Environmental changes caused by climate change influence food and water supplies. Rising temperatures have changed precipitation patterns, and severe weather events are altering ecosystems, reducing agricultural production, and worsening water shortages (Loucks, 2021). Such as:

Agriculture Adaptation: Growing climate-resistant crops reduces crop failures and ensures food security. This can require choosing heat-, drought-, and flood-tolerant crops. Rainwater collection, drip irrigation, and recycling can preserve water and lessen agricultural systems' drought sensitivity. Protecting soil health, reducing erosion, and improving water infiltration can boost agricultural resistance to climate change. Creating severe weather warning systems can help farmers adapt to climate change (Singh et al., 2022).

Water Resource Adaptation: Infrastructure modifications, including dams, reservoirs, and irrigation systems, can increase water storage and delivery, lowering climate change susceptibility. Water conservation, such as minimizing home and industrial water waste, can assist in ensuring a sustainable water supply. Considering the connectivity of water, land, and ecosystems, integrated water resource management (IWRM) can help manage water resources more efficiently (Du et al., 2022).

Ecosystem-Based Adaptation: Restoring damaged forests, wetlands, and coastal ecosystems can increase climate change resistance and offer ecosystem services like water filtering and flood control. Biodiversity conservation can improve ecosystem resilience and lessen climate change vulnerability of food and water systems. This includes protecting endangered species, creating protected areas, and supporting sustainable land use (Tabari, 2020).

Governance and Policy: Climate change adaptation requires good policy and governance. Governments can encourage climate-resilient activities, fund research, and create thorough adaptation plans. Addressing climate change and sharing information and best practices requires international collaboration (Ray & Majumder, 2024).

Therefore, sustainable food and water security requires climate change resilience. Effective adaptation techniques can reduce climate change's adverse effects on food and water systems and promote a sustainable future.

2.7.4 Policy Coherence for Food and Water Sustainability

Sustainable food and water management requires policy consistency. Governments can foster sustainable development by aligning and complementing policies across sectors (Galli et al., 2020). Such as:

Agriculture, Water, Environment, and Climate Change: Food, water, Environment, and climate change are linked; thus, policymaking must be comprehensive. Ensuring these policies function together for sustainable development requires alignment. Sustainable farming requires water, environmental, and climate change policies for appropriate water availability and quality, ecosystem protection, and greenhouse gas reduction (Singh & Nath, 2020).

Integration and Coordination of Policies: Policy integration unites policy sectors to accomplish goals. This can be done via cross-sectoral planning, coordination, and integrated policy frameworks. By combining policies, governments can avoid competing goals and promote mutual reinforcement (Nhamo et al., 2022).

Institutional Structure: Effective institutions are needed to execute clear policies. This can entail founding or strengthening institutions. Institutions facilitating coordination, information exchange, and decision-making can help in policy coherence (Beecher, 2020).

Engagement of Stakeholders: Policy formulation and implementation must include many stakeholders to ensure relevance, effectiveness, and equity. Governments can obtain insights and tailor policies to local requirements by incorporating farmers, communities, companies, and other stakeholders (Sohrabi et al., 2022).

Monitoring and Assessing: Monitoring and evaluating policies regularly helps determine their efficacy and opportunities for improvement. Governments can change policies to promote sustainable food and water management by measuring progress and recognizing obstacles.

International Cooperation: Global food and water sustainability requires international collaboration. International agreements and organizations can encourage policy consistency, knowledge and technology transfer, and transboundary water challenges (Anser et al., 2020).

Sustainable food and water management requires policy consistency. Governments can foster sustainable development by aligning and complementing policies across sectors. Policy coherence and sustainable food and water management need policy integration, institutional frameworks, stakeholder participation, monitoring and evaluation, and international collaboration (Oguh et al., 2021).

2.8 Future Research Direction

This thorough report covers food and water sustainability issues and prospects. However, new challenges need further investigation. Future research includes quantifying interdependencies, studying climate change impacts and adaptation, investing in sustainable food and water management technology, analyzing policies, implementing community-based approaches, addressing equity and justice, evaluating international cooperation, and staying current on emerging issues. Quantifying food, water, and environmental interdependencies requires establishing models to reliably measure them, examining trade-offs and synergies at various scales, and quantifying resource management techniques' economic and ecological costs and benefits. Innovative local adaptation methods must be developed and evaluated for climate change effects and adaptability. Existing policies, new policies, and community-based alternatives should be analyzed and implemented.

2.9 Conclusion

This extensive research examines the complex link between food, water, and the Environment and the obstacles to sustainable growth. The study identifies population increase, urbanization, climate change, and environmental degradation as major problems. These issues worsen water shortages, reduced agricultural output, and food poverty. These difficulties need new solutions, including technical advances, legislative changes, and community-based methods. The study has demonstrated how integrated water resource management (IWRM) is essential for sustainable water usage and fair distribution, and agroforestry, organic farming, and conservation agriculture can boost food production and conserve the Environment. Climate change adaptation measures are crucial to prevent negative consequences on food and water systems. Policy coherence promotes sustainable food and water management by harmonizing policies across sectors. It has shown how community-based initiatives allow local communities to manage resources sustainably and promote food security. According to the study, food and water sustainability needs technology advances, regulatory changes, and community-based efforts. Understanding food, water, and the Environment and adopting successful solutions can lead to sustainable development and a secure and fair future for future generations.

References

Abou-Shady, A., Siddique, M. S., & Yu, W. (2023). A critical review of innovations and perspectives for providing adequate water for sustainable irrigation. *Water, 15*(17), 3023.

Agustang, A., Adam, A., & Upe, A. (2021). Community empowerment strategy towards a sustainable rural community-based tourism village. *Turkish Online Journal of Qualitative Inquiry, 12*(3).

Amanullah, Khalid, S., Imran, Khan, H. A., Arif, M., Altawaha, A. R., et al. (2020). Effects of climate change on irrigation water quality. In *Environment, climate, plant and vegetation growth* (pp. 123–132).

Ameen, M., Zafar, A., Javaid, M. M., Zia, M. A., Mahmood, A., Naqve, M., & Bibi, S. (2023). Climate-resilient technology for maize production. In *Climate-resilient agriculture, Vol 2: Agro-biotechnological advancement for crop production* (pp. 157–188). Springer International Publishing.

Amri, S., & Sihotang, J. (2023). Impact of poverty reduction programs on healthcare access in remote areas: Fostering community development for sustainable health. *Law and Economics, 17*(3), 170–185.

Anser, M. K., Yousaf, Z., Usman, B., Nassani, A. A., Abro, M. M. Q., & Zaman, K. (2020). Management of water, energy, and food resources: Go for green policies. *Journal of Cleaner Production, 251*, Article 119662.

Ashraf, S. A., Siddiqui, A. J., Abd Elmoneim, O. E., Khan, M. I., Patel, M., Alreshidi, M., et al. (2021). Innovations in nanoscience for the sustainable development of food and agriculture with implications on health and environment. *Science of the Total Environment, 768*, Article 144990.

Avgoustaki, D. D., & Xydis, G. (2020). How energy innovation in indoor vertical farming can improve food security, sustainability, and food safety? In *Advances in food security and sustainability* (Vol. 5, pp. 1–51). Elsevier.

Beecher, J. A. (2020). Policy note: A universal equity–efficiency model for pricing water. *Water Economics and Policy, 6*(03), 2071001.

Benson, D., Gain, A. K., & Giupponi, C. (2020). Moving beyond water centricity? Conceptualizing integrated water resources management for implementing sustainable development goals. *Sustainability Science, 15*(2), 671–681.

Beretić, N., Bauer, A., Funaro, M., Spano, D., & Marras, S. (2024). A participatory framework to evaluate coherence between climate change adaptation and sustainable development policies. *Environmental Policy and Governance, 34*(3), 275–290.

Berry, E. M. (2020). Food insecurity, social inequity, and sustainability. *Hidden Hunger and the Transformation of Food Systems, 121*, 95–104.

Birkmann, J., Liwenga, E., Pandey, R., Boyd, E., Djalante, R., Gemenne, F., et al. (2022). Poverty, livelihoods and sustainable development.

Bolwig, S., Tanner, A. N., Riemann, P., Redlingshöfer, B., & Zhang, Y. (2021). Reducing consumer food waste using green and digital technologies.

Borsellino, V., Schimmenti, E., & El Bilali, H. (2020). Agri-food markets towards sustainable patterns. *Sustainability, 12*(6), 2193.

Bradu, P., Biswas, A., Nair, C., Sreevalsakumar, S., Patil, M., Kannampuzha, S., et al. (2023). Recent advances in green technology and Industrial Revolution 4.0 for a sustainable future. *Environmental Science and Pollution Research, 30*(60), 124488–124519.

Salmi, J., & D'Addio, A. (2021). Policies for achieving inclusion in higher education. *Policy Reviews in Higher Education, 5*(1), 47–72.

Scharlemann, J. P., Brock, R. C., Balfour, N., Brown, C., Burgess, N. D., Guth, M. K., et al. (2020). Towards understanding interactions between Sustainable Development Goals: The role of environment–human linkages. *Sustainability Science, 15*, 1573–1584.

Shah, M. A., Kumar, S., & Talukder, M. B. (2024). Advancing to sustainable development: A comprehensive analysis of climate change, green energy, green investment, and research trends. In *AI applications for clean energy and sustainability* (pp. 1–15). IGI Global.

Sherwood, J. (2020). The significance of biomass in a circular economy. *Bioresource Technology, 300*, Article 122755.

Siegel, K. M., & Lima, M. G. B. (2020). When international sustainability frameworks encounter domestic politics: The sustainable development goals and agri-food governance in South America. *World Development, 135*, Article 105053.

Sigalla, O. Z., Tumbo, M., & Joseph, J. (2021). Multi-stakeholder platform in water resources management: A critical analysis of stakeholders' participation for sustainable water resources. *Sustainability, 13*(16), 9260.

Singh, C., & Nath, R. (2020). *Farming system and sustainable agriculture: Agricultural reform.* Sgoc Publication.

Singh, R. P., Qidwai, S., Singh, O., Reddy, B. R., Saharan, S., Kataria, S. K., et al. (2022). Millets for food and nutritional security in the context of climate resilient agriculture: A review. *International Journal of Plant & Soil Science*, 939–953.

Skawińska, E., & Zalewski, R. I. (2022). Combining the water–energy–food and food waste–food loss–food security nexuses to reduce resource waste. *Energies, 15*(16), 5866.

Skendžić, S., Zovko, M., Živković, I. P., Lešić, V., & Lemić, D. (2021). The impact of climate change on agricultural insect pests. *Insects, 12*(5), 440.

Smith, V. H., & Glauber, J. W. (2020). Trade, policy, and food security. *Agricultural Economics, 51*(1), 159–171.

Sohrabi, M., Ahani Amineh, Z. B., Niksokhan, M. H., & Zanjanian, H. (2022). A framework for optimal water allocation considering water value, strategic management and conflict resolution. *Environment, Development and Sustainability*, 1–32.

Srivastav, A. L., Dhyani, R., Ranjan, M., Madhav, S., & Sillanpää, M. (2021). Climate-resilient strategies for sustainable management of water resources and agriculture. *Environmental Science and Pollution Research, 28*(31), 41576–41595.

Tabari, H. (2020). Climate change impact on flood and extreme precipitation increases with water availability. *Scientific Reports, 10*(1), 13768.

Talukder, M. B., Khan, M. R., & Kumar, S. (2024a). Socioeconomic gaps and foster inclusive growth: sustainable tourism initiatives. In *Building community resiliency and sustainability with tourism development* (pp. 56–82). IGI Global.

Talukder, M. B., Kumar, S., & Das, I. R. (2024b). Food wastage on the economic outcome: Evidence from the hotel industry. In *Sustainable disposal methods of food wastes in hospitality operations* (pp. 68–80). IGI Global.

Talukder, M. B., Kumar, S., & Khan, M. R. (2024c). The Integration of sustainable practices in cultural and gastronomic tourism. In *Contemporary solutions for sustainable transportation practices* (pp. 467–484). IGI Global.

Tambovceva, T. T., Melnyk, L. H., Dehtyarova, I. B., & Nikolaev, S. O. (2021). Circular economy: Tendencies and development perspectives.

Upadhyay, R. K. (2020). Markers for global climate change and its impact on social, biological and ecological systems: A review. *American Journal of Climate Change, 9*(03), 159.

Upadhyaya, A. (2020). Water conservation and management technologies for its efficient utilization. *Journal of Natural Resource Conservation and Management, 1*(1), 20–28.

van Ginkel, M., & Biradar, C. (2021). Drought early warning in agri-food systems. *Climate, 9*(9), 134.

Vanham, D., & Leip, A. (2020). Sustainable food system policies need to address environmental pressures and impacts: The example of water use and water stress. *Science of the Total Environment, 730*, Article 139151.

Varma, V. G. (2022). Water-efficient technologies for sustainable development. In *Current directions in water scarcity research* (Vol. 6, pp. 101–128). Elsevier.

Venkatesharaju, K. (2022). Water resources and its need in environmental science. *Message from the Editor-in-Chief*.

Wang, X. C., Jiang, P., Yang, L., Van Fan, Y., Klemeš, J. J., & Wang, Y. (2021). Extended water-energy nexus contribution to environmentally-related sustainable development goals. *Renewable and Sustainable Energy Reviews, 150*, Article 111485.

Wato, T., Amare, M., Bonga, E., Demand, B. B. O., & Coalition, B. B. R. (2020). The agricultural water pollution and its minimization strategies—A review. *Journal of Resources Development and Management, 64*, 10–22.

Wikantiyoso, R., Cahyaningsih, D. S., Sulaksono, A. G., Widayati, S., Poerwoningsih, D., & Triyosoputri, E. (2021). Development of sustainable community-based tourism in Kampong Grangsil, Jambangan Village, Dampit District, Malang Regency. *International Review for Spatial Planning and Sustainable Development, 9*(1), 64–77.

Yue, Q., Wu, H., Wang, Y., & Guo, P. (2021). Achieving sustainable development goals in agricultural energy-water-food nexus system: An integrated inexact multi-objective optimization approach. *Resources, Conservation and Recycling, 174*, Article 105833.

Zahedi, R., Yousefi, H., Aslani, A., & Ahmadi, R. (2024). Water, energy, food and environment nexus (WEFEN): Sustainable transition, gaps and Covering approaches. *Energy Strategy Reviews, 54*, Article 101496.

Zahoor, I., & Mushtaq, A. (2023). Water pollution from agricultural activities: A critical global review. *International Journal of Chemical and Biochemical Sciences, 23*(1), 164–176.

Zakar, M. Z., Zakar, D. R., & Fischer, F. (2020). Climate change-induced water scarcity: A threat to human health. *South Asian Studies, 27*(2).

Zelber-Sagi, S., Carrieri, P., Pericàs, J. M., Ivancovsky-Wajcman, D., Younossi, Z. M., & Lazarus, J. V. (2024). Food inequity and insecurity and MASLD: Burden, challenges, and interventions. *Nature Reviews Gastroenterology & Hepatology*, 1–19.

Zeng, Y., Liu, D., Guo, S., Xiong, L., Liu, P., Yin, J., & Wu, Z. (2022). A system dynamic model to quantify the impacts of water resources allocation on water–energy–food–society (WEFS) nexus. *Hydrology and Earth System Sciences, 26*(15), 3965–3988.

Zhang, D., Sial, M. S., Ahmad, N., Filipe, A. J., Thu, P. A., Zia-Ud-Din, M., & Caleiro, A. B. (2020). Water scarcity and sustainability in an emerging economy: A management perspective for future. *Sustainability, 13*(1), 144.

Zohuri, B. (2023). Navigating the global energy landscape balancing growth, demand, and sustainability. *Journal of Material Sciences & Applied Engineering, 2*(7).

Chapter 3
Food and Water Sustainability: Contribution Toward SDGs

Chakshu Gupta, Raman Jasrotia, and Seema Langer

Abstract The sustainability of food and water systems is critical to achieving the United Nations' Sustainable Development Goals (SDGs), particularly SDG 2 (Zero Hunger) and SDG 6 (Clean Water and Sanitation). This work examines the complex interdependencies between food production and water resource management, emphasizing sustainable practices that address global challenges like water scarcity, food insecurity, and environmental degradation. As population growth, climate change, and resource constraints intensify, efficient management of these systems is essential for long-term resilience. Key sustainability strategies explored include precision agriculture, agroecology, and integrated water resource management (IWRM), which enhance food security and water conservation while contributing to other SDGs such as SDG 1 (No Poverty), SDG 3 (Good Health and Well-Being), SDG 12 (Responsible Consumption and Production), and SDG 13 (Climate Action). Innovative technologies, such as sensor-based irrigation, artificial intelligence, and wastewater recycling, are highlighted as transformative tools for advancing sustainability in agriculture and water systems. Case studies from water-scarce regions illustrate the successful adoption of these technologies, with a focus on community-driven initiatives and supportive policy frameworks. The critical role of women, especially in developing regions, is emphasized as a key factor in effective resource management. Furthermore, the analysis explores the role of governance structures, policies, and financing mechanisms, such as green bonds and public–private partnerships, in promoting sustainable food and water systems. By integrating circular economy principles and developing climate-resilient crops, this work outlines future directions for mitigating climate change impacts and conserving resources. Ultimately, it calls for global cooperation and action from policymakers, businesses, and communities to adopt sustainable practices, ensuring a secure and resilient future for all.

C. Gupta · R. Jasrotia (✉) · S. Langer
Department of Zoology, University of Jammu, Jammu, Jammu and Kashmir, India
e-mail: drraman.jasrotia@jammuuniversity.ac.in

R. Jasrotia
Department of Zoology, Cluster University of Jammu, Jammu, Jammu and Kashmir, India

J. Parray and A. K. Haghi (eds.), *Food and Water Security*,
https://doi.org/10.1007/978-3-032-04870-7_3

Keywords Agroecology · Circular economy · Drip irrigation · SDG · Sustainability · Recycling · Waste water

3.1 Introduction

Sustainability in food and water systems refers to development and management practices that ensure the long-term availability, quality, and efficient use of essential resources. In these systems, sustainability embodies the balance between meeting present-day needs and conserving resources for future generations (FAO, 2017). It requires addressing environmental, economic, and social dimensions simultaneously, which includes minimizing water and soil pollution, ensuring food security, and improving resilience to climate change impacts (Rockström et al., 2009). Sustainable practices aim to optimize resource use while reducing negative environmental impacts, such as water waste, over-extraction, soil degradation, and biodiversity loss (Pretty & Bharucha, 2014).

Water sustainability involves managing freshwater resources to maintain ecosystem health and human use, while food sustainability emphasizes efficient agricultural practices, the reduction of food loss, and enhancing nutritional outcomes (Molden, 2007). These two sectors are deeply interdependent—water scarcity directly affects agricultural output, while unsustainable food production practices can degrade water quality and availability (Vörösmarty et al., 2010). Consequently, sustainable food and water systems are pivotal in shaping a secure and equitable future.

The United Nations' 17 Sustainable Development Goals (SDGs) provide a global blueprint for a sustainable future, aimed at ending poverty, protecting the planet, and ensuring prosperity for all by 2030 (United Nations, 2015). Among these, SDG 2 (Zero Hunger) and SDG 6 (Clean Water and Sanitation) are directly related to food and water sustainability, focusing on ending hunger, achieving food security, improving nutrition, promoting sustainable agriculture, and ensuring the availability and sustainable management of water and sanitation for all (Uhlenbrook & Connor, 2019).

SDG 2 seeks to eliminate hunger by promoting sustainable agricultural practices that increase productivity while preserving ecosystems and ensuring food security for growing populations (Clapp, 2020). SDG 6 emphasizes the need for equitable access to clean water and adequate sanitation, particularly in regions facing water scarcity or contamination (Huns, 2018). Both goals intersect with numerous other SDGs, such as SDG 1 (No Poverty), SDG 3 (Good Health and Well-being), SDG 12 (Responsible Consumption and Production), and SDG 13 (Climate Action), reinforcing the interconnectedness of sustainable development (United Nations, 2015). Food and water systems are fundamental to achieving global sustainability. The availability and quality of these resources underpin human health, economic development, and environmental sustainability. Food security, which hinges on sustainable agricultural practices, is crucial for poverty alleviation and improved health outcomes (SDG 1

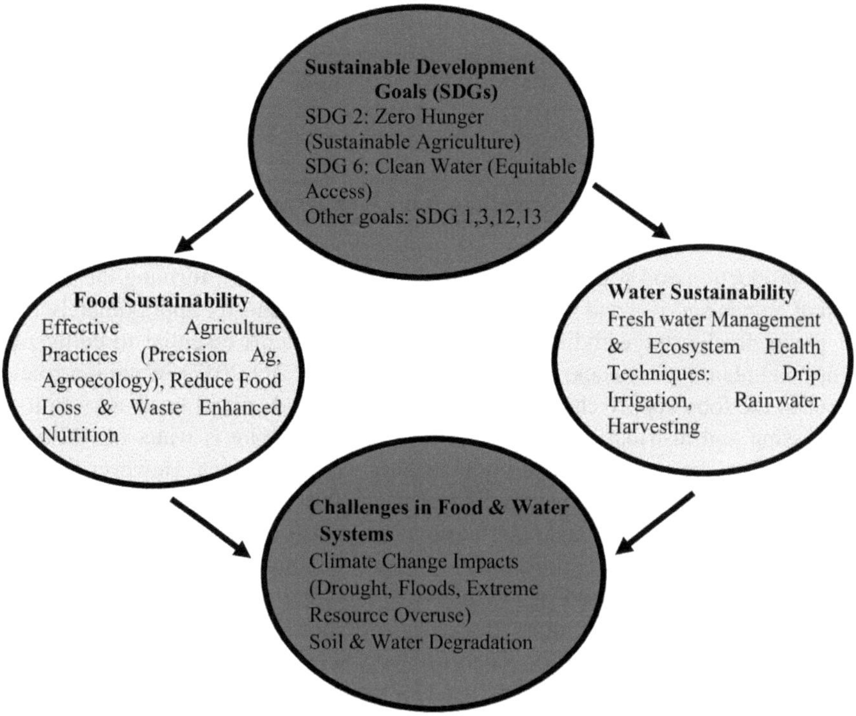

Fig. 3.1 Conceptual diagram of the interconnectedness of food and water sustainability, challenges, and the Sustainable Development Goals (SDGs)

and SDG 3) (Pretty et al., 2018). Sustainable water management is essential not only for direct human consumption but also for agriculture, industry, and the preservation of aquatic ecosystems (Molden, 2007) (Fig. 3.1).

Addressing sustainability in these systems requires a comprehensive understanding of their environmental, social, and economic dimensions. For example, agriculture is responsible for approximately 70% of global freshwater withdrawals (FAO, 2017), making it one of the most water-intensive industries. Water scarcity currently affects 40% of the global population (Mekonnen & Hoekstra, 2016), with projections suggesting that by 2050, more than half of the world's population could be living in water-stressed regions (Uhlenbrook & Connor, 2019). This highlights the need for sustainable water use strategies in agricultural practices, such as drip irrigation, rainwater harvesting, and wastewater recycling (Grafton et al., 2018).

By adopting sustainable food production methods like precision agriculture, agroecology, and integrated water resource management (IWRM), countries can boost food security while conserving water and maintaining biodiversity (Pretty et al., 2018). Such approaches are also essential to mitigating the effects of climate change, which threatens to exacerbate food and water insecurity in many vulnerable regions (Change, 2013). As climate change intensifies droughts, floods, and extreme weather events, the pressure on global food and water systems is increasing, necessitating a stronger focus on resilience-building strategies in these sectors (Field et al., 2014) (Fig. 3.1).

3.2 Food and Water Systems: An Overview

Food production and water resources are inextricably linked, forming the foundation of human survival and well-being. Agriculture accounts for approximately 70% of global freshwater withdrawals, with irrigation systems essential to cultivating crops, maintaining livestock, and supporting fisheries (FAO, 2017). Water is a crucial input in the food supply chain, from crop irrigation and animal husbandry to food processing and distribution (Molden, 2007). As agriculture is water-intensive, its sustainability depends on efficiently managing water resources. However, agricultural practices can also negatively impact water systems. Excessive use of chemical fertilizers, pesticides, and poor waste management can lead to water pollution, affecting both surface and groundwater quality (Vörösmarty et al., 2010).

The connection between water systems and food production is even more evident in regions where water scarcity threatens agricultural productivity. In arid and semi-arid areas, competition for water resources between agricultural use and other sectors is increasing, highlighting the need for integrated water resource management (IWRM) approaches to ensure water allocation that supports both food security and ecosystem health (Pretty et al., 2018). The health of aquatic ecosystems also directly impacts fisheries and aquaculture, which are key sources of food and livelihood for many communities worldwide (FAO, 2020).

The growing global population, projected to reach nearly 10 billion by 2050, places unprecedented pressure on food and water systems (UN, 2019). Population growth demands more food production, while simultaneously, climate change exacerbates water scarcity, extreme weather events, and changes in precipitation patterns, all of which threaten agricultural yields and water availability (Field et al., 2014). Additionally, around 2.3 billion people currently live in water-stressed regions, and this figure is expected to rise as climate change impacts intensify (Uhlenbrook & Connor, 2019).

Water scarcity is often compounded by the overexploitation of water resources, degradation of natural water bodies, and poor water management practices (Mekonnen & Hoekstra, 2016). Food insecurity is another major global challenge, with approximately 690 million people suffering from hunger in 2019, a number that has only increased with recent global events like the COVID-19 pandemic (WHO, 2020). Environmental degradation, including deforestation, soil erosion, and biodiversity loss, further threatens the sustainability of food and water systems. These global challenges underscore the urgency of transitioning toward sustainable food and water management strategies that promote resilience and adaptability.

Sustainability in food and water systems centers around key principles aimed at fostering resilience, reducing environmental impacts, and ensuring equitable access to resources. "Resource efficiency" is critical in addressing water and food challenges, ensuring that agricultural practices minimize water use and reduce wastage across the supply chain (Rockström et al., 2009). Practices like "precision agriculture"—which uses technologies such as sensor-based irrigation systems to optimize water usage—and agroecological approaches help achieve greater productivity with fewer inputs (Pretty et al., 2018).

Reducing the "environmental footprint" of food production is another core principle. This involves lowering greenhouse gas emissions, reducing chemical runoff, and preventing water pollution by adopting sustainable farming techniques, promoting organic farming, and protecting natural water bodies (Molden, 2007). "Biodiversity enhancement" is equally important, as biodiversity underpins the resilience of both food and water systems. Diversified crops, sustainable livestock management, and protection of aquatic biodiversity help strengthen ecosystems, making them more resistant to climate-induced stresses (IPBES, 2019).

Ensuring "equitable access" to food and water resources is vital for achieving sustainability. This requires policies and frameworks that prioritize vulnerable populations, enhance rural livelihoods, and improve food distribution networks, while promoting water access for all through IWRM (Grafton et al., 2018). The promotion of sustainable diets, reduction of food loss and waste, and innovation in water-saving technologies are crucial steps toward ensuring food and water security for a growing population (FAO, 2017).

3.3 Contributions of Food and Water Sustainability to SDG 2: Zero Hunger

Water security is intrinsically linked to food security, as water is a critical input for agricultural production, which provides the majority of the world's food supply. Sustainable water management is essential in areas prone to droughts and water scarcity, where unreliable access to water severely limits agricultural productivity. In sub-Saharan Africa and South Asia, regions that already face food insecurity, unpredictable rainfall and depleted water resources exacerbate the vulnerability of smallholder farmers (WHO, 2020). In such regions, effective water management through rainwater harvesting, drip irrigation, and groundwater recharge can significantly improve food production and bolster food security (Foley et al., 2011). The implementation of integrated water resource management (IWRM) ensures the efficient allocation of water across competing sectors while simultaneously protecting aquatic ecosystems essential for fisheries and agriculture (Grafton et al., 2018).

In regions where food insecurity is driven by water shortages, ensuring equitable water distribution and promoting policies that prioritize agricultural water use can have a profound impact on reducing hunger. Agricultural water productivity, which measures the amount of food produced per unit of water, is a key indicator of the sustainability of food systems (Molden, 2007). Thus, sustainable water management not only supports SDG 6 (Clean Water and Sanitation) but also plays a pivotal role in achieving SDG 2 (Zero Hunger).

Sustainable agricultural practices are crucial for increasing food production while minimizing water consumption, contributing directly to SDG 2. Precision agriculture, which utilizes technologies such as soil moisture sensors and satellite data, allows farmers to optimize water use by targeting irrigation precisely when and where it is needed (Rockström et al., 2009). This practice reduces water wastage and increases crop yields, particularly in water-scarce regions. Agroecology, another sustainable approach, promotes agricultural systems that work in harmony with the environment. By enhancing biodiversity, maintaining soil fertility, and conserving water through natural means, agroecological methods improve the resilience of food systems to climate change and water scarcity (Altieri & Nicholls, 2017).

Permaculture and organic farming further promote sustainability by focusing on holistic land management practices. Organic farming reduces the reliance on chemical fertilizers and pesticides, which often contribute to water pollution, thereby improving both soil and water health (IFOAM, 2017). These sustainable practices not only ensure food security by increasing yields but also protect critical water resources essential for long-term agricultural productivity.

Sustainability in food and water systems is not only about increasing yields but also about enhancing the quality of food. Ensuring access to nutrient-rich food is a key aspect of addressing global hunger and malnutrition. Sustainable water management directly influences the diversity of agricultural outputs, as crops with higher nutritional value, such as fruits, vegetables, and legumes, often require more precise water control compared to staple crops (FAO, 2017). Diverse cropping systems that include a range of nutrient-rich crops can improve dietary diversity, particularly in regions where malnutrition is prevalent (Hodgkin et al., 2015).

Additionally, initiatives like food fortification and biofortification—where crops are bred to contain higher levels of essential nutrients—can further enhance food security by providing populations with access to more nutritious food, reducing dependence on water-intensive crops like rice and wheat (Bouis & Saltzman, 2017). By improving water use efficiency in agriculture, farmers can diversify their crops, offering a wider array of nutrient-rich foods to their communities and contributing to both food security and better health outcomes, aligning with SDG 3 (Good Health and Well-Being).

Minimizing food waste is critical to achieving both SDG 2 and SDG 12 (Responsible Consumption and Production), and it also has significant implications for water sustainability. Food waste accounts for a substantial portion of water usage, as water is embedded in all stages of food production, from cultivation to consumption. Globally, approximately one-third of all food produced is lost or wasted, representing not just a loss of food but also a massive amount of wasted water (Gustavsson et al., 2011). Reducing food waste across the supply chain—from production and harvesting to retail and household consumption—can save vast quantities of water that would otherwise go unused (Mekonnen & Hoekstra, 2011).

Implementing strategies to prevent food loss at the farm level, such as improving storage facilities, transportation infrastructure, and market access, can reduce post-harvest losses, particularly in developing countries. In more developed regions, educating consumers on proper food storage and waste minimization techniques can significantly reduce household food waste. Reducing food waste also lightens the environmental burden on water resources, as less water is required to produce food that will not be discarded, thereby contributing to the sustainability of both food and water systems.

3.4 Contributions of Food and Water Sustainability to SDG 6: Clean Water and Sanitation

Agriculture accounts for 70% of global freshwater withdrawals, making it the largest consumer of water. Efficient water use in agriculture is crucial for the sustainability of both food and water systems, especially in regions facing water scarcity. Techniques such as drip irrigation, which delivers water directly to plant roots, can significantly improve water-use efficiency, reducing water loss through evaporation and runoff (Postel et al., 2001). Similarly, rainwater harvesting—capturing and storing rain for agricultural use—can help mitigate the impact of erratic rainfall and reduce dependence on traditional water sources (Ngigi, 2003). In addition, wastewater reuse in agriculture offers a sustainable alternative for irrigation, particularly in water-scarce regions, by treating and safely using domestic or industrial wastewater for crop production (Qadir et al., 2010). These practices not only conserve water but also enhance agricultural resilience to climate change.

Agricultural activities can have a significant impact on water quality, particularly through runoff containing fertilizers, pesticides, and sediments. The adoption of sustainable farming practices, such as reduced chemical inputs, organic farming, and agroforestry, can help minimize water pollution and protect water resources (Pretty et al., 2018). The excessive use of nitrogen and phosphorus in agriculture often leads to the eutrophication of nearby water bodies, causing algal blooms and negatively affecting aquatic ecosystems (Tilman et al., 2001). Therefore, integrated nutrient management and agroecological practices that reduce dependency on chemical fertilizers are essential for maintaining water quality. Furthermore, buffer strips of vegetation along waterways and constructed wetlands can help filter out pollutants before they reach water bodies, playing a critical role in reducing agricultural water contamination (Dosskey et al., 2010).

Integrated Water Resource Management (IWRM) offers a holistic approach to managing water resources by promoting the coordinated development of water, land, and related resources to maximize economic and social welfare without compromising the sustainability of ecosystems (TAC, 2000). In the context of agriculture, IWRM ensures that water is allocated equitably across sectors, including domestic, agricultural, and industrial users while maintaining ecological health. By incorporating the principles of IWRM, countries can address competing demands for water and prevent conflicts over water allocation, especially in water-scarce regions (Giordano & Shah, 2017). Additionally, IWRM promotes the integration of ecosystem services into water management plans, ensuring that water use in agriculture does not compromise the health of aquatic ecosystems, which are crucial for supporting biodiversity and food systems (Marlow et al., 2010).

Community-led initiatives are essential for achieving SDG 6, particularly in low-resource settings. These projects often integrate sustainable water use and sanitation into food systems by involving local stakeholders in the management of water resources. For example, the Self-Employed Women's Association (SEWA) in India has developed community-managed water and sanitation systems that ensure both reliable water supply and improved agricultural productivity (Panda, 2005). These projects often incorporate rainwater harvesting, low-cost sanitation solutions, and sustainable irrigation methods, demonstrating that integrating food and water sustainability into local governance can lead to long-term improvements in both water access and food security. Such community-driven approaches empower local populations to take ownership of their water resources, leading to more effective and sustainable solutions to water and sanitation challenges.

3.5 Contributions to Other SDGs

3.5.1 SDG 1: No Poverty

Sustainable food and water systems are pivotal for poverty reduction, especially in rural and marginalized communities. These systems support income generation, job creation, and economic stability, offering a pathway out of poverty. Sustainable agricultural practices, such as agroecological methods, provide consistent income opportunities for smallholder farmers, promoting local employment and economic resilience. Furthermore, by stabilizing food prices and improving access to clean water, these practices reduce the burden of waterborne diseases and enhance living standards. The socio-economic benefits that sustainable food and water systems bring to impoverished communities include reducing dependency on external aid and fostering self-sufficiency (Fan & Brzeska, 2016; Fanzo et al., 2020).

3.5.2 SDG 3: Good Health and Well-Being

Access to clean water and nutritious food is fundamental to public health, reducing malnutrition, improving sanitation, and decreasing waterborne diseases. Sustainable systems enhance food quality and variety, which helps address malnutrition and supports immune health. They also limit the need for harmful chemicals, benefiting soil and water quality and improving food safety. Sustainable food systems increase the availability of nutrient-rich foods, and improved water management reduces the prevalence of diarrheal diseases, both contributing to healthier communities (Black et al., 2013; Willett et al., 2019).

3.5.3 SDG 7: Affordable and Clean Energy

Water is essential in clean energy production, particularly for hydropower. Sustainable water practices, including efficient irrigation and water recycling, support energy sustainability by conserving water resources needed for energy production. Additionally, renewable energy solutions in water management reduce greenhouse gas emissions and enhance energy access for remote areas. Integrated water-energy systems play a critical role in supporting sustainable energy production and affordable access, especially in energy-scarce regions (Poff et al., 2015; Rasul & Sharma, 2015).

3.5.4 SDG 12: Responsible Consumption and Production

Promoting responsible consumption and production in food and water systems conserves resources and reduces waste. Sustainable diets—particularly those with plant-based foods—have a lower environmental impact and reduce water demand. Water-efficient practices like drip irrigation and rainwater harvesting help conserve water resources while reducing food loss preserves both water and energy in the production chain. The sustainable consumption patterns are crucial to resource conservation and minimizing the ecological footprint of food systems (Kummu et al., 2012; Tilman & Clark, 2014).

3.5.5 SDG 13: Climate Action

Food systems and water resources are closely linked to climate resilience. Sustainable agricultural practices, such as agroforestry and conservation tillage, help mitigate climate impacts by enhancing soil water retention and reducing greenhouse gas emissions. These practices build resilience against climate variability, helping communities manage water scarcity and droughts. Climate-smart agriculture and sustainable water management are essential for protecting water resources and reducing the climate footprint of agriculture, supporting climate adaptation and mitigation efforts (Campbell et al., 2017; Lal, 2015).

3.6 Case Studies: Food and Water Sustainability in Action

3.6.1 Sustainable Farming in Water-Scarce Regions

In arid and semi-arid regions, innovative farming practices have proven essential for sustaining agriculture despite limited water resources. Israel, a leader in water-efficient agriculture, has pioneered drip irrigation technology, which delivers water directly to plant roots with minimal evaporation loss, reducing water usage by up to 50% compared to traditional methods (Frankel, 2021). In India, research has promoted practices like the System of Rice Intensification (SRI), a water-efficient method for growing rice that decreases water use by approximately 30% while boosting yield (Thakur et al., 2014). These approaches are valuable examples for other water-scarce areas, as they help enhance crop productivity while conserving limited water resources. Sustainable water management strategies like these contribute directly to SDG 6 (Clean Water and Sanitation) and SDG 2 (Zero Hunger) by promoting resilient agricultural practices that support food security.

3.6.2 Water-Saving Technologies in Agriculture

Advances in technology play a crucial role in enhancing water sustainability in agriculture. Sensor-based irrigation systems, for instance, allow precise water application based on real-time soil moisture data, reducing water use and improving crop health (Jones, 2004). Additionally, desalination technology has enabled water-stressed regions, such as the Middle East, to transform seawater into a viable resource for agriculture, although it remains costly and energy-intensive (Ghaffour et al., 2013). Smart water management platforms, which incorporate IoT (Internet of Things) and AI, enable data-driven water use monitoring, providing insights for optimized irrigation schedules and improved resource efficiency (Arpula et al., 2024). These innovations are critical for achieving SDG 12 (Responsible Consumption and Production) by ensuring that agricultural water usage is efficient, sustainable, and capable of meeting future demands.

3.6.3 Urban Agriculture and Water Efficiency

Urban agriculture, including vertical farming, hydroponics, and aquaponics, represents a significant step toward sustainable food production within city environments, minimizing the environmental footprint of transporting food from rural areas. Vertical farms can produce large quantities of food on small urban plots using controlled environments that require significantly less water than traditional farming; hydroponics, for instance, uses up to 90% less water (Al-Kodmany, 2018). Aquaponics systems, which combine fish farming with hydroponics, recycle water between plants and fish, creating a closed-loop system that enhances water efficiency and reduces waste (Love et al., 2015). Urban farming initiatives address SDG 11 (Sustainable Cities and Communities) by integrating food production into urban planning and reducing resource strain on rural farmland, thereby supporting local food security and sustainability.

3.6.4 Role of Women in Food and Water Sustainability

Women play a central role in managing food production and water resources, especially in rural areas of developing nations. Studies show that women are primarily responsible for agricultural activities and water management at the household and community levels, yet they often face limited access to resources and decision-making power (Johnson et al., 2016). Empowering women through education, resources, and support can significantly boost sustainability outcomes. For instance, programs that provide women with training in sustainable agriculture techniques have demonstrated improvements in crop yield, soil health, and water use efficiency (Meinzen-Dick et al., 2011). Empowering women aligns with SDG 5 (Gender Equality) and contributes to SDGs 2 and 6 by enhancing resource management and fostering more resilient food systems.

3.7 Policy and Governance for Food and Water Sustainability

3.7.1 International and National Policies on Sustainable Food and Water

International and national policies provide a framework for promoting food and water sustainability. Key international agreements, such as the Paris Agreement, underscore the urgent need for climate resilience in food and water systems by setting targets to reduce greenhouse gas emissions, which directly impact agriculture and water resources (Strohmaier et al., 2016). The UN Decade on Water for Sustainable Development (2018–2028) has further highlighted the need for integrated water resource management to address global water scarcity, advocating for the implementation of sustainable water practices to secure reliable water supplies for food production (UN, 2018). National policies often align with these global initiatives; for example, the European Union's Common Agricultural Policy (CAP) encourages member states to adopt sustainable agriculture practices that minimize water pollution and promote biodiversity (Pe'er et al., 2017). Such frameworks are instrumental in aligning local actions with global sustainability goals, contributing to SDG 2 (Zero Hunger) and SDG 6 (Clean Water and Sanitation).

3.7.2 Regulatory Frameworks for Sustainable Agriculture

At the national level, regulatory frameworks play a crucial role in encouraging sustainable agricultural practices and conserving water resources. Countries like the United States have developed programs under the Farm Bill, such as the Environmental Quality Incentives Program (EQIP), which offers financial incentives to farmers who adopt practices that improve soil health and water conservation (Claassen et al., 2018). Similarly, the European Union's Water Framework Directive establishes strict guidelines on water management to prevent over-extraction and pollution, aiming to protect aquatic ecosystems while ensuring sustainable agricultural water use. Regulatory policies like these incentivize farmers to adopt conservation practices that reduce water usage, thereby supporting SDG 12 (Responsible Consumption and Production) and enhancing ecosystem resilience.

3.7.3 Financing Sustainable Food and Water Projects

Innovative financing mechanisms have emerged as key tools in supporting sustainable food and water initiatives. Green bonds, for instance, are increasingly used by governments and corporations to fund projects that promote sustainable land use, water conservation, and climate resilience (Maltais & Nykvist, 2020). Water funds, which pool financial resources from various stakeholders to restore and protect watersheds, have shown success in regions like Latin America by providing stable water supplies for agriculture while preserving natural ecosystems (Abell et al., 2017). Public–private partnerships (PPPs) also facilitate resource sharing, enabling large-scale investments in infrastructure and technology for sustainable food and water systems, especially in developing countries with limited funding for sustainability projects (Petersen-Perlman et al., 2017). Such financing models are essential for scaling up sustainable practices, thereby contributing to SDG 13 (Climate Action) and fostering long-term food and water security.

3.7.4 Global Cooperation on Water and Food Security

Cross-border cooperation on water resources management and food security is essential, particularly in regions with shared water bodies, such as the Nile, Mekong, and Ganges rivers. Transboundary water agreements, like the Nile Basin Initiative, help countries collaboratively manage shared water resources to avoid conflicts and ensure equitable distribution for agricultural and domestic use (Paisley & Henshaw, 2013). In the Mekong region, multilateral efforts such as the Mekong River Commission (MRC) facilitate data-sharing and joint planning among riparian countries to ensure sustainable water management and food security (Sithirith, 2021). Cross-border cooperation enhances regional resilience to water scarcity, supports sustainable agricultural practices, and reduces the risk of geopolitical tensions, aligning with SDG 6 (Clean Water and Sanitation) and SDG 2 (Zero Hunger) by promoting a collaborative approach to resource management.

3.8 The Role of Technology in Advancing Food and Water Sustainability

3.8.1 Innovative Technologies for Water Conservation

Innovative technologies like nanotechnology, artificial intelligence (AI), big data, and remote sensing are transforming water conservation efforts. For example, nanotechnology enables the creation of highly efficient filtration systems, helping to purify water with minimal resource use, which is particularly beneficial in water-scarce regions (Shannon et al., 2008). AI and machine learning are being applied to monitor water usage, predict water needs, and detect leaks in real-time, reducing water waste at large scales (Krishnan et al., 2022). Big data analytics and remote sensing allow for detailed mapping of water resources and agricultural areas, providing valuable insights into water stress and optimizing irrigation practices. These advancements contribute directly to SDG 6 (Clean Water and Sanitation) by improving water-use efficiency and supporting sustainable water management across agriculture and urban settings.

3.8.2 Precision Agriculture for Sustainable Food Production

Precision agriculture is a transformative approach that uses drones, sensors, and satellite data to monitor crops and manage resources efficiently. By providing precise information on soil moisture, nutrient levels, and crop health, these technologies allow farmers to apply inputs like water and fertilizers exactly where needed, reducing waste and enhancing productivity (Liakos et al., 2018). Drones and field sensors can assess crop stress levels and detect early signs of diseases, facilitating timely intervention and improving yield. Precision agriculture not only optimizes water use but also reduces the environmental impact of farming, supporting SDG 2 (Zero Hunger) by ensuring food security while conserving essential resources.

3.8.3 Water Treatment and Recycling Technologies

Water treatment, desalination, and wastewater recycling technologies are essential for sustainable water management in agriculture and urban areas. Advances in membrane filtration, for example, have improved desalination efficiency, making it a more viable option for freshwater production in arid regions (Elimelech & Phillip, 2011). Wastewater recycling technologies, including membrane bioreactors, convert wastewater into a reusable resource, thus reducing dependency on freshwater sources and enabling sustainable irrigation (Hai et al., 2018). These technologies align with SDG 6 (Clean Water and Sanitation) and SDG 12 (Responsible Consumption and Production) by enhancing water availability and supporting circular water management practices in both urban and agricultural landscapes.

3.8.4 Digital Platforms for Food-Water Nexus Management

Digital platforms that integrate data on food production, water use, and environmental conditions are emerging as powerful tools for managing the food-water nexus sustainably. Platforms such as Agri-Tech and Water-Energy-Food Nexus apps use data integration and predictive analytics to provide stakeholders with real-time insights into resource availability and demands, enabling more efficient planning and allocation (Bazilian et al., 2011). These platforms support collaboration across sectors by facilitating data-sharing and decision-making, which is critical for balancing food production with water conservation. By advancing digital infrastructure for resource monitoring and management, these platforms contribute to SDG 12 (Responsible Consumption and Production) and SDG 13 (Climate Action), helping create resilient, sustainable systems for food and water security.

3.9 Future Directions and Emerging Trends in Food and Water Sustainability

3.9.1 Sustainable Diets and Water Conservation

As concerns about resource scarcity and environmental degradation rise, there is an increasing global shift toward plant-based diets, which have a significantly lower water footprint compared to animal-based diets. Research shows that plant-based foods generally require less water to produce; for example, producing a kilogram of vegetables requires significantly less water than the same quantity of beef or dairy products (Mekonnen & Hoekstra, 2012). Adopting plant-based diets supports both SDG 6 (Clean Water and Sanitation) by conserving water resources and SDG 12 (Responsible Consumption and Production) by encouraging sustainable food choices. This dietary shift is gaining momentum through public awareness and policy initiatives, which promote the environmental and health benefits of reduced meat consumption.

3.9.2 Climate-Resilient Crops

Advances in agricultural biotechnology are paving the way for climate-resilient crops that are drought-tolerant and water-efficient. Through genetic modification and traditional breeding techniques, scientists are developing crops that can withstand extreme weather and utilize water more efficiently, thereby enhancing food security in water-scarce regions (Nelson et al., 2018). These innovations support SDG 2 (Zero Hunger) by ensuring that crops can thrive despite climate change pressures. Drought-resistant crops like genetically modified maize and salt-tolerant rice varieties help sustain agricultural productivity while reducing the need for excessive irrigation, demonstrating how technology and plant science can directly contribute to sustainable food and water systems.

3.9.3 Circular Economy in Food and Water Systems

The principles of a circular economy are being increasingly integrated into food and water systems to minimize waste and maximize resource efficiency. By focusing on reuse, recycling, and recovery, circular economy models can reduce the environmental footprint of food production and water usage. For instance, organic waste from food production can be repurposed as biofertilizer, while wastewater can be treated and reused for irrigation. These approaches not only decrease the reliance on freshwater resources but also support sustainable agriculture by recycling nutrients within the system (Kirchherr et al., 2017). Implementing circular economy practices aligns with SDG 12 (Responsible Consumption and Production) and promotes a holistic, resource-efficient approach to food and water sustainability.

3.10 Future Policy Directions

Achieving long-term sustainability in food and water systems will require continued evolution in global governance, international cooperation, and policy frameworks. Future policies are expected to address resource distribution more equitably, promote cross-border water-sharing agreements, and incentivize sustainable agricultural practices worldwide (Sandberg et al., 2019). International collaborations, like the Paris Agreement and the UN Decade on Water for Sustainable Development, highlight the importance of policy synergy to address the complex, interlinked challenges of food and water security. As climate pressures intensify, proactive policies and binding international agreements will be crucial for achieving the Sustainable Development Goals, particularly those related to food security, water access, and ecosystem resilience.

3.11 Conclusion

Food and water sustainability is fundamental to achieving multiple Sustainable Development Goals (SDGs), particularly those focused on poverty reduction, health, climate action, and responsible consumption. Sustainable practices in food and water management not only enhance resource efficiency but also promote environmental health, economic stability, and social equity. As explored in this chapter, food and water systems are intricately linked; sustainable practices in one area significantly impact outcomes in the other, demonstrating the interdependence of resources critical for human well-being and ecosystem resilience (Rockström et al., 2009). Addressing food and water sustainability holistically is thus essential for progressing toward the SDGs and creating resilient communities.

Policymakers, businesses, and local communities must collaborate to advance sustainable practices in food and water systems. Policymakers should prioritize legislation that incentivizes conservation and supports sustainable agricultural methods, while businesses play a crucial role in innovating eco-friendly practices across supply chains. Meanwhile, community involvement is vital for the successful implementation of water-saving and resource-efficient practices at a local level. Achieving the SDGs by 2030 requires immediate action and commitment from all sectors to drive the transition toward sustainability and ensure these critical resources are available for future generations (United Nations, 2015).

Looking forward, the future of food and water sustainability will be shaped by both challenges and opportunities. Climate change, population growth, and resource scarcity will test the resilience of food and water systems, making innovation and technological advancement essential for adaptation. Emerging technologies in water conservation, precision agriculture, and resource recycling offer promising solutions but require investment and international cooperation to scale effectively (Foley et al., 2011). Collaboration across borders will be increasingly necessary to address shared water resources and cross-border food trade sustainably. Thus, innovation, collective action, and a commitment to resilience will be central to overcoming future obstacles and ensuring the long-term sustainability of food and water systems.

References

Abell, R., Asquith, N., Boccaletti, G., Bremer, L., Chapin, E., Erickson-Quiroz, A., Higgins, J., Johnson, J., Kang, S., Karres, N., & Lehner, B. (2017). *Beyond the source: The environmental, economic and community benefits of source water protection.* The Nature Conservancy.
Al-Kodmany, K. (2018). The vertical farm: A review of developments and implications for the vertical city. *Buildings, 8*(2), 24.
Altieri, M. A., & Nicholls, C. I. (2017). Agroecology: A brief account of its origins and currents of thought in Latin America. *Agroecology and Sustainable Food Systems, 41*(3–4), 231–237.
Arpula, S., Gouthami, B., Nayak, B. B., & Shirisha, K. (2024). *Smart irrigation systems in agriculture. Insights into agricultural sciences* (Vol. II).
Bazilian, M., Rogner, H., Howells, M., Hermann, S., Arent, D., Gielen, D., Steduto, P., Mueller, A., Komor, P., Tol, R. S., & Yumkella, K. K. (2011). Considering the energy, water and food nexus: Towards an integrated modelling approach. *Energy Policy, 39*(12), 7896–7906.
Black, R. E., Victora, C. G., Walker, S. P., Bhutta, Z. A., Christian, P., De Onis, M., Ezzati, M., Grantham-McGregor, S., Katz, J., Martorell, R., & Uauy, R. (2013). Maternal and child undernutrition and overweight in low-income and middle-income countries. *The Lancet, 382*(9890), 427–451.
Bouis, H. E., & Saltzman, A. (2017). Improving nutrition through biofortification: A review of evidence from HarvestPlus, 2003 through 2016. *Global Food Security, 12*, 49–58.
Campbell, B. M., Beare, D. J., Bennett, E. M., Hall-Spencer, J. M., Ingram, J. S., Jaramillo, F., Ortiz, R., Ramankutty, N., Sayer, J A., & Shindell, D. (2017). Agriculture production as a major driver of the Earth system exceeding planetary boundaries. *Ecology and Society, 22*(4).
Change, IPCC Climate. (2013). The physical science basis.
Claassen, R., Bowman, M., McFadden, J., Smith, D., & Wallander, S. (2018). *Tillage intensity and conservation cropping in the United States.* USDA ERS.
Clapp, J. (2020). Food security and nutrition: Building a global narrative towards 2030.
Dosskey, M. G., Vidon, P., Gurwick, N. P., Allan, C. J., Duval, T. P., & Lowrance, R. (2010). The role of riparian vegetation in protecting and improving chemical water quality in streams 1. *JAWRA, 46*(2), 261–277.
Elimelech, M., & Phillip, W. A. (2011). The future of seawater desalination: Energy, technology, and the environment. *Science, 333*(6043), 712–717.
Fan, S., & Brzeska, J. (2016). Sustainable food security and nutrition: Demystifying conventional beliefs. *Global Food Security, 11.*
Fanzo, J., Haddad, L., McLaren, R., Marshall, Q., Davis, C., Herforth, A., Jones, A., Beal, T., Tschirley, D., Bellows, A., & Miachon, L. (2020). The Food Systems Dashboard is a new tool to inform better food policy. *Nature Food, 1*(5), 243–246.
FAO. (2020). *The state of world fisheries and aquaculture 2020. Sustainability in action.* Rome.
Field, C. B., Barros, V. R., Dokken, D. J., Mach, K. J., Mastrandrea, M. D., Bilir, T. E., Chatterjee, M., Ebi, K. L., Estrada, Y. O., Genova, R. C., Girma, B., Kissel, E. S., Levy, A. N., MacCracken, S., Mastrandrea, P. R., & White, L. L. (2014). Climate change 2014: impacts, adaptation, and vulnerability. In *Part A: Global and sectoral aspects. Contribution of working group II to the fifth assessment report of the intergovernmental panel on climate change.* Cambridge University Press, Cambridge, United Kingdom and New York, NY, USA.
Foley, J. A., Ramankutty, N., Brauman, K. A., Cassidy, E. S., Gerber, J. S., Johnston, M., Mueller, N. D., O'Connell, C., Ray, D. K., West, P. C., & Balzer, C. (2011). Solutions for a cultivated planet. *Nature, 478*(7369), 337–342.
Food and Agriculture Organization of the United Nations. (2017). *The future of food and agriculture: Trends and challenges.* FAO.
Frankel, D. (2021). Drip irrigation: Israel's ingenious invention. Hasbara fellowships.
Ghaffour, N., Missimer, T. M., & Amy, G. L. (2013). Technical review and evaluation of the economics of water desalination: Current and future challenges for better water supply sustainability. *Desalination, 309*, 197–207.

Giordano, M., & Shah, T. (2017). From IWRM back to integrated water resources management. In *Revisiting integrated water resources management* (pp. 4–16). Routledge.
Grafton, R. Q., Williams, J., Perry, C. J., Molle, F., Ringler, C., Steduto, P., Udall, B., Wheeler, S. A., Wang, Y., Garrick, D., & Allen, R. G. (2018). The paradox of irrigation efficiency. *Science, 361*(6404), 748–750.
Gustavsson, J., Cederberg, C., Sonesson, U., & van Otterdijk, M. A. (2011). *Global food losses and food waste: Extent, causes and prevention.* FAO.
Hai, F. I., Yamamoto, K., & Lee, C. H. (Eds.) (2018). *Membrane biological reactors: Theory, modeling, design, management and applications to wastewater reuse.* Iwa Publishing.
Hodgkin, T., Hunterm D., Wood, S., & Demers, N. R. (2015). *Agricultural biodiversity, food security and human health.* UNEP, Convention on Biological Diversity, WHO.
Huns, P. (2018) "Nature-based solutions" and global water shortages: A political ecology of The United Nation's world water development report.
IFOAM (2017) *The IFOAM norms for organic production and processing: Version 2014.* International Federation of Organic Agriculture Movements.
IPBES W. (2019) *Intergovernmental science-policy platform on biodiversity and ecosystem services. Summary for Policy Makers of the Global Assessment Report on Biodiversity and Ecosystem Services of the Intergovernmental Science-Policy Platform on Biodiversity and Ecosystem Services.* IPBES Secretariat.
Johnson, N. L., Kovarik, C., Meinzen-Dick, R., Njuki, J., & Quisumbing, A. (2016). Gender, assets, and agricultural development: Lessons from eight projects. *World Development, 83*, 295–311.
Jones, H. G. (2004). Irrigation scheduling: Advantages and pitfalls of plant-based methods. *Journal of Experimental Botany, 55*(407), 2427–2436.
Kirchherr, J., Reike, D., & Hekkert, M. (2017). Conceptualizing the circular economy: An analysis of 114 definitions. *RCR, 127*, 221–232.
Krishnan, S. R., Nallakaruppan, M. K., Chengoden, R., Koppu, S., Iyapparaja, M., Sadhasivam, J., & Sethuraman, S. (2022). Smart water resource management using Artificial Intelligence—A review. *Sustainability, 14*(20), 13384.
Kummu, M., De Moel, H., Porkka, M., Siebert, S., Varis, O., & Ward, P. J. (2012). Lost food, wasted resources: Global food supply chain losses and their impacts on freshwater, cropland, and fertiliser use. *Science of the Total Environment, 438*, 477–489.
Lal, R. (2015). Soil carbon sequestration in agroecosystems of India. *JISSS, 63*(2), 125.
Liakos, K. G., Busato, P., Moshou, D., Pearson, S., & Bochtis, D. (2018). Machine learning in agriculture: A review. *Sensors, 18*(8), 2674.
Love, D. C., Fry, J. P., Li, X., Hill, E. S., Genello, L., Semmens, K., & Thompson, R. E. (2015). Commercial aquaponics production and profitability: Findings from an international survey. *Aquaculture, 435*, 67–74.
Maltais, A., & Nykvist, B. (2020). Understanding the role of green bonds in advancing sustainability. *JSF&I*, 1–20.
Marlow, D. R., Beale, D. J., & Burn, S. (2010). A pathway to a more sustainable water sector: Sustainability-based asset management. *Water Science and Technology, 61*(5), 1245–1255.
Meinzen-Dick, R., Quisumbing, A., Behrman, J., Biermayr-Jenzano, P., Wilde, V., Noordeloos, M., Ragasa, C., & Beintema, N. (2011) *Engendering agricultural research, development and extension.* Intl Food Policy Res Inst.
Mekonnen, M. M., & Hoekstra, A. Y. (2011). The green, blue and grey water footprint of crops and derived crop products. *HESS, 15*(5), 1577–1600.
Mekonnen, M. M., & Hoekstra, A. Y. (2012). A global assessment of the water footprint of farm animal products. *Ecosystems, 15*(3), 401–415.
Mekonnen, M. M., & Hoekstra, A. Y. (2016). Four billion people facing severe water scarcity. *Science Advances, 2*(2), Article e1500323.
Molden, D. (2007). *Water for food water for life: A comprehensive assessment of water management in agriculture.* Routledge.

Nelson, G., Bogard, J., Lividini, K., Arsenault, J., Riley, M., Sulser, T. B., Mason-D'Croz, D., Power, B., Gustafson, D., Herrero, M., & Wiebe, K. (2018). Income growth and climate change effects on global nutrition security to mid-century. *Nature Sustainability, 1*(12), 773–781.

Ngigi, S. N. (2003). What is the limit of up-scaling rainwater harvesting in a river basin? *Physics and Chemistry of the Earth, Parts A/B/C, 28*(20–27), 943–956.

Paisley, R. K., & Henshaw, T. W. (2013). Transboundary governance of the Nile River Basin: Past, present and future. *Environmental Development, 7*, 59–71.

Panda, S. (2005). Women's role in local water management: Insights from SEWA's millennium water campaign, Gujarat (India). https://doi.org/10.13140/2.1.2916.3529

Pe'er, G., Lakner, S., Müller, R., Passoni, G., Bontzorlos, V., Clough, D., Moreira, F., Azam, C., Berger, J., Bezak, P., & Bonn, A. (2017). *Is the CAP Fit for purpose? An evidence-based fitness-check assessment.* German Centre for Integrative Biodiversity Research (iDiv) Halle-Jena-Leipzig.

Petersen-Perlman, J. D., Veilleux, J. C., & Wolf, A. T. (2017). International water conflict and cooperation: Challenges and opportunities. *Water International, 42*(2), 105–120.

Poff, N., Brown, C., Grantham, T., Mendoza, G., Dominique, K., Haasnoot, M., Baeza, A., Spence, C., Wilby, R., Palmer, M., & Matthews, J. (2015). Sustainable water management under future uncertainty with eco-engineering decision scaling. *Nature Climate Change.*

Postel, S., Polak, P., Gonzales, F., & Keller, J. (2001). Drip irrigation for small farmers: A new initiative to alleviate hunger and poverty. *Water International, 26*(1), 3–13.

Pretty, J., & Bharucha, Z. P. (2014). Sustainable intensification in agricultural systems. *Annals of Botany, 1:114*(8):1571–1596.

Pretty, J., Benton, T. G., Bharucha, Z. P., Dicks, L. V., Flora, C. B., Godfray, H. C., Goulson, D., Hartley, S., Lampkin, N., Morris, C., & Pierzynski, G. (2018). Global assessment of agricultural system redesign for sustainable intensification. *Nature Sustainability, 1*(8), 441–446.

Qadir, M., Wichelns, D., Raschid-Sally, L., McCornick, P. G., Drechsel, P., Bahri, A., & Minhas, P. S. (2010). The challenges of wastewater irrigation in developing countries. *Agricultural Water Management, 97*(4), 561–568.

Rasul, G., & Sharma, B. (2015). The nexus approach to water–energy–food security: An option for adaptation to climate change. *Climate Policy, 16*(6), 682–702.

Rockström, J., Steffen, W., Noone, K., Persson, Å., Chapin, F. S., Lambin, E. F., Lenton, T. M., Scheffer, M., Folke, C., Schellnhuber, H. J., & Nykvist, B. (2009). A safe operating space for humanity. *Nature, 461*(7263), 472–475.

Sandberg, M., Klockars, K., & Wilén, K. (2019). Green growth or degrowth? Assessing the normative justifications for environmental sustainability and economic growth through critical social theory. *Journal of Cleaner Production, 206*, 133–141.

Shannon, M. A., Bohn, P. W., Elimelech, M., Georgiadis, J. G., Mariñas, B. J., & Mayes, A. M. (2008). Science and technology for water purification in the coming decades. *Nature, 452*(7185), 301–310.

Sithirith, M. (2021). Transboundary river basin governance: A case of the Mekong River Basin. In *River basin management-sustainability issues and planning strategies*

Strohmaier, R., Rioux, J., Seggel, A., Meybeck, A., Bernoux, M., Salvatore, M., Miranda, J., & Agostini, A. (2016). The agriculture sectors in the Intended Nationally Determined Contributions: Analysis. Environment and Natural Resources Management Working Paper 62.

TAC G. (2000). Integrated water resources management. TAC Background Paper No. 4. Global Water Partnership Technical Advisory Committee, Stockholm.

Thakur, A., Mohanty, C., Raychaudhuri, S., Verma, O., Panda, & Kumar, A. (2014) Integrated system of rice intensification (SRI) for enhancing land and water productivity. Research bulletin no. 70, ICAR, Odisha, India.

Tilman, D., & Clark, M. (2014). Global diets link environmental sustainability and human health. *Nature, 515*(7528), 518–522.

Tilman, D., Fargione, J., Wolff, B., D'antonio, C., Dobson, A., Howarth, R., Schindler, D., Schlesinger, W. H., Simberloff, D., & Swackhamer, D. (2001). Forecasting agriculturally driven global environmental change. *Science, 292*(5515), 281–284.

Uhlenbrook, S., & Connor, R. (2019). The United Nations world water development report 2019: Leaving no one behind.

United Nations. (2015). *Transforming our world: The 2030 agenda for Sustainable Development United Nations.*

United Nations. (2018). *International decade for action 'Water for Sustainable Development' 2018–2028.* UNGA.

United Nations. (2019). *World population prospects 2024.* Department of Economic and Social Affairs, Population Division.

Vörösmarty, C. J., McIntyre, P. B., Gessner, M. O., Dudgeon, D., Prusevich, A., Green, P., Glidden, S., Bunn, S. E., Sullivan, C. A., Liermann, C. R., & Davies, P. (2010). Global threats to human water security and river biodiversity. *Nature, 467*(7315), 555–561.

Willett, W., Rockström, J., Loken, B., Springmann, M., Lang, T., Vermeulen, S., Garnett, T., Tilman, D., DeClerck, F., Wood, A., & Jonell, M. (2019). Food in the Anthropocene: The EAT-Lancet Commission on healthy diets from sustainable food systems. *The Lancet, 393*(10170), 447–492.

World Health Organization. (2020). *The state of food security and nutrition in the world 2020: Transforming food systems for affordable healthy diets.* Food & Agriculture Org.

Part II
Key Challenges and Threats

Chapter 4
Climate Change and the Challenges of Water and Food Security

Shahid Gul, Shakir Ahmad Mir, and Yaqoota Maqbool

Abstract Climate change is a Universal concern. There is hardly any aspect of life that climate change has not influenced, be it humans, the environment they live in or the world economy. This chapter examines the interconnected relationship between water scarcity, water quality, food security and their dependence on climate change across the globe. Particularly in the scope of this chapter, valuable response options such as drip irrigation, planting climate-smart crops, and adopting agroforestry are appraised. The chapter also underscores the need for mitigation, including agricultural emission reductions and conversion of energy services to renewables. Addressing the complex water-food nexus in climate stress calls for a participatory, integrated framework. In this context, this chapter aims to constitute policy recommendations that enhance sustainable management of water and food resources in climate change by presenting examples of good practices and discussing the constraints to implementing those practices.

Keywords Climate change · Water scarcity · Food insecurity · Asia · Sustainability · Africa · Water-food nexus · Energy services

4.1 Introduction

There are so many pressing issues that humanity is facing today. Climate change is one such issue. Climate change refers to a significant change or shift concerning different aspects such as temperature, precipitation, and other factors that can substantially be linked to weather conditions (Abbass et al., 2022). The interference of human activity primarily causes it. With the coming of human civilization, industries came into being and agriculture was recognized. Since then, massive deforestation has occurred due

S. Gul (✉) · S. A. Mir
ICAR - Central Institute of Fisheries Education, Mumbai, India
e-mail: Shahidgul9616@gmail.com

Y. Maqbool
College of Fisheries, Lembucherra, Tripura, India

J. Parray and A. K. Haghi (eds.), *Food and Water Security*,
https://doi.org/10.1007/978-3-032-04870-7_4

to rising populations and greenhouse gas emissions from industrial processes. These gases have detrimental effects on the environment, and the degree of the greenhouse effect has only increased over the years. As a result, global temperatures are now approximately 1.1° (1.8 °F) above pre-industrial levels, leading to climate change (IPCC, 2021).

Due to this altercation graph, the Earth has undergone severe climatic changes in the last hundred years, which is the reason behind unprecedented changes in human behaviors. Global warming has caused significant climatic disruptions, including rising sea levels, frequent heatwaves, droughts, floods, and hurricanes that never used to occur at this frequency or in this much severity before industrialization (USEPA, 2024). Such catastrophes negatively impact ecosystems and greatly affect societies' functioning, especially in developing countries with limited resources. Predictions before climate change were catastrophic, and researchers contended that the temperatures would get warmer than they already are. In about 50 or 100 years, we could be looking at Global temperatures increasing to somewhere between 1.5 and 2 °C, which is a threshold not easily recoverable. This is especially catastrophic, as this will result in rising sea levels, which can engulf entire coastlines, leading to internal displacement of humans (Hoegh et al., 2018).

While the climate crisis is becoming more and more severe with every passing moment, it has consequences for many sectors but not many sectors are as exposed as water and food security.

4.2 Water and Food Security: A Foundation for Human Livelihoods

Water and food security are key to ensuring life can be supported on Earth. Water security is the confidence in having enough freshwater resources of acceptable quality that are adequate for human, ecosystem, and economic activities' requirements. In the same way, food security, as envisaged by the United Nations, exists when people have access to enough food that is safe and nutritious to lead an active and healthy life at all times.

Water and food security are also important to sustainable livelihoods, particularly in rural areas where agriculture, fisheries and aquaculture activities are the mainstay of the local economies. These sectors depend on water resources not only for irrigation water but also for domestic, sanitary and industrial uses. With the ever-increasing world population predicted to reach 9.8 million by 2050, the demand for water and food will be proportionately high (UN DESA, 2024). In turn, this makes it even more important to deal with issues related to climate change, which will likely increase the prevalence of water and food insecurity worldwide.

Human health also requires water and food security. The availability of clean water is crucial to maintaining good sanitary and hygienic practices and avoiding diseases caused by polluted water. In addition, adequate food security is relevant to the fight against undernourishment, which is still a global challenge (Ebi et al., 2021). Insufficient availability of these resources has economic ramifications, as water-related diseases and poor nutrition impair economic productivity and foster poverty.

Moreover, the nexus of water and food security is not one-way; food systems are also influenced by water resource availability and vice versa. When water becomes scarcer due to climate change, food production is constrained, causing volatility and a shortage of food prices. The need of the hour is to ensure that there is a consistent supply of both resources, and this will require the adoption of sustainable management practices, appropriate development of new technologies and coordinated global responses.

4.3 Impact of Climate Change on Water Security

i. **Changes in Water Availability: Droughts and Floods**
   Changes in precipitation patterns are the most noticeable effects of climate change on the water security of the affected areas. There is a notable trend in the global rainfall patterns wherein secondary areas witness heightened drought and flooding conditions. The dry and semi-dry regions are growing drier in climate, so freshwater sources are cut short for drinking and industrial use (Pizzorni et al., 2024). This is the same case for low-lying coastal areas at a higher risk of destructive flooding since they are infested by severe and more frequent storms. Longer droughts are occurring, and the severity of droughts is increasing, affecting the water supply for many people. Based on various examples, Sub-Saharan Africa, Mediterranean regions, and parts of Australia are experiencing prolonged dry periods affecting water supply and agricultural output (Xu et al., 2023). Increasing temperature means accelerating the evaporation rate from lakes and rivers, further worsening the water crisis.
   On the other hand, flooding has also increased in some regions, which may destroy freshwater sources. Due to the increase in global temperature, there is more rainfall, which causes massive floods, damages infrastructure, displaces people and destroys crops. These floods can often result in waterlogged conditions, which can deteriorate water resources, cause the spread of water-borne diseases, and scarcity of clean water (Fig. 4.1).
ii. **Decreasing Water Quality**
   In addition to the concern of water scarcity, global warming also affects the quality of water resources. The increased heat is also accompanied by an increase in the amount of rainfall channeled through runoff. The runoff brings pollutants like pesticides, fertilizers, and sewage into rivers and lakes, and these water sources are then used for drinking. In addition, warm water promotes harmful algal blooms, which destroy water quality and are dangerous to living beings that reside in water. Besides, in coastal areas, rising sea levels and droughts

Fig. 4.1 Projected impacts of climate change on food and water security in various countries FAO (2021), UNEP (2023, 2024), UNEPA (2024)

cause more intrusion of salt water into the freshwater aquifers, which lowers the amount of good water surrounding areas have (Dao et al., 2024).

iii. **Groundwater Depletion**
More and more regions switch to groundwater use to deal with the erratic character of surface water sources. Still, relying on groundwater to meet the needs of irrigation, industry, or domestic purposes is destructive in most cases, especially in regions with prolonged droughts. When groundwater supplies are used up, years and even decades are required to restore them through natural processes. Such depletion is a cause for concern when one considers the future availability of freshwater in regions with already limited water supplies, as is the case with a large part of India, the United States, and the Middle East, among other regions (Bhattarai et al., 2023).

4.4 Impact of Climate Change on Food Security

i. **Reduced Agricultural Productivity**
Climate change affects agricultural output straightforwardly through changes in weather conditions, increases in average temperature and extreme weather (Zhou et al., 2024). Temperature-sensitive crops such as maize, wheat, and rice are very affected. Increased average temperatures lead to shorter growing periods, lower agricultural production and greater chances of crop diseases and pests. The variations and deforestation of exploitable rainfall patterns add another layer of difficulty on top of crop planning since some areas may experience droughts and others too much rainfall and flooding. Such changes can bring down global food production, in the most extreme case, by 25% by 2050 (Bouteska et al., 2024).
In addition to crop productivity, livestock production is also at risk. Higher temperatures increase the risk of heat stress in animals, reducing reproductive rates, milk production, and overall health. Additionally, the availability of grazing land is threatened by desertification and changing vegetation patterns, especially in arid regions.

ii. **Fisheries and Aquaculture**
The problem of fishery management and the future sustainability of fishery resources is continuously worsened globally due to climate change impacts on water temperature, ocean acidification, and the rising sea level. Temperature increase influences fish species' migratory behavior and breeding patterns in oceans, while an increase of CO_2 in oceans results in ocean acidification, causing damage to reefs and many marine environments. Temperatures continuing to increase also threaten freshwater fish species, as the warming of rivers and lakes diminishes their oxygen reserves and adds to the pressure of global stress on aquatic ecosystems (Cheung et al., 2021). A great number of people are dependent on aquaculture resources for their sustenance and income. Therefore,

communities' dependent on the fishery for food sources and income are, in turn, suffering from great challenges when it comes to this sector (Gul et al., 2024).

iii. **Disruptions of the Food Supply Chain**
Hurricanes, floods, droughts, and other climate-related disasters have also had irreversible effects on food supply chains. Extreme weather can impair transportation systems, storage units, and other food processing factories and, in turn, result in food loss with less accessibility to it. Furthermore, extreme weather leads to agricultural disruption that handles food. When this happens, food prices tend to increase, worsening food insecurity, especially for vulnerable groups in developing nations.

4.5 The Water-Food Nexus: Interconnectedness and Challenges

Water and food systems are closely linked, as food production in agriculture, aquaculture, and food processing all require a precious resource: water. Mixed systems in developed countries where modern inputs are used have a lot of advantages, as agricultural production in such conditions does not suffer considerably from water restrictions. However, in developing countries and regions, a low level of modernization and adaptation to dry agriculture conditions makes it unsustainable. Collectively, agriculture uses up to 70% of the world's available freshwater resources (Smith et al., 2023).

The relations between water and food security are alarming in regions and countries with limited water. For example, in many developing nations, where water-dependent crops are widely cultivated, agricultural output capabilities are critically impaired by falling water resources over time. In this case, a suitable response that addresses water and food security must be integrated, and general resource challenges must be appropriately addressed.

4.5.1 Regional Variations and Inequalities

SPIs' (Southern Pakistan India's climate) impacts on water and food security are not the same everywhere. Resources, especially those in warm and subtropical areas, are seriously disadvantaged (Baocheng et al., 2024). For instance, these Kibera slum areas of Sub-Saharan Africa are losing their once reliable rain for crop farming, a major food producer. Similarly, parts of South Asia that depend on the Himalayan range glaciers for crops are also affected by the receding glaciers, posing a more adverse effect (Shah & Lele, 2011).

Small Island Developing States (SIDS) are also extremely vulnerable as they suffer from consequences such as rising sea levels, saltwater intrusion, and diminishing arable land. At the same time, tropical storms further worsen the conditions for coastal populations. These countries are usually poor in infrastructure, technology and finances to adjust to these changes and, hence, are more vulnerable to food and water insecurity (Vousdoukas et al., 2023).

4.6 Adaptation Strategies for Water and Food Security

Therefore, since the risk caused by climate change is rising, effective adaptation mechanisms should be adopted to secure water and food. Adaptation means adjusting systems and practices to decrease the susceptibility to climate change impacts and increase the capacity to withstand the change. In this context, this sub-chapter discusses sustainable water resource management, integrated agricultural systems capable of responding to climate variability, and food system strengthening measures against climate variability (Table 4.1).

Table 4.1 Strategies and benefits for adapting to climate change

Adaptation strategy	Implementation level	Benefits	Example	References
Rooftop rainwater harvesting	Household/ community	Improves water availability across the year	Rooftop gardening and harvesting in urban India	Puppala et al. (2023)
Drought-resistant crop varieties	Agro-horticulture/ farm	Ensures stable crop production under severe drought conditions	Drought-tolerant rice in Asia and Africa	Fisher et al. (2015), Dar et al. (2020), Adjah et al. (2022)
Early warning mechanisms	State/country/ regional/global	Reduces loss of life and property	Cyclone warnings in Southeast Asia	Sharma et al. (2009), Tran et al. (2022)
Cold storage facilities	Agriculture and fisheries value chain	Reduces wastage of food	Cold chains management in horticulture and fisheries sector in southeast Asia and east Africa	Han et al. (2021), Makule et al. (2022), Mustafa et al. (2024)
Agroforestry approaches	Community/ regional	Enhances water and soil retention capacity	Tree integration in European and Kenyan farms	Rois-Diaz et al. (2018), Nyaga et al. (2019), Kitonga et al. (2020)

4.6.1 Sustainable Water Management Practices

Water management practices can mitigate water security issues that arise as a result of climate change. The main aim of sustainable water management approaches is to enhance efficiency. Some of the practices include the following:

i. **Efficient Irrigation Systems**: Water and water resources management are vital in this regard as agriculture is the most popular water-consuming activity on a global scale, and improved irrigation systems will be essential for climate change adaptation. Traditional irrigation methods, such as flood irrigation, often waste large amounts of water. In contrast, modern systems like drip irrigation deliver water directly to plant roots, minimizing evaporation and runoff. This system is particularly effective in water-scarce regions, conserving water and energy.
ii. **Rainwater Harvesting**: This practice is primitive yet remains one of the most effective ways of solving the water crisis. To at least have enough water during the dry seasons, people collect rainwater in the rainy season. Rooftop rainwater harvesting systems are especially relevant in urban areas as they reduce the need for municipal water systems. In large-scale harvesting systems, such areas provide safe and ample water for irrigation, drinking, and sanitation.
iii. **Groundwater Recharge**: In places where a large proportion of freshwater is sourced from groundwater, artificial recharge systems are necessary. Overused aquifers can be effectively replenished with structures like check dams, recharge wells, and percolation tanks, providing viable solutions for the long term. Such approaches are particularly relevant in India's arid states, where groundwater depletion is much higher than replenishments.
iv. **Climate-Resilient Agriculture**: Due to climate change and its effect on the climate, which is exacerbating global warming further, the need to shift agricultural practices has become paramount. Emphasis must be placed on agricultural techniques that can withstand extreme weather events, droughts and pests. Climate-resilient agriculture lies on the pillars of ensuring no-till farming to maximize overall yields with minimal environmental costs.
v. **Drought-Tolerant Crops**: Research on cultivating drought-tolerant and heat-resistant crops has grown in importance and relevance. These crops are relatively less water-demanding, can sustain higher temperatures and yield under hostile climatic environments. For instance, drought-tolerant maize and salt-resistant rice have been developed and implemented in water-scarce regions and those prone to salinization, including Sub-Saharan Africa and the coastal regions of South Asia.
vi. **Agroforestry**: Agroforestry is the practice of incorporating trees into agricultural systems and landscapes, which has advantages, including increasing biodiversity, improving the soil's health and fertility, and increasing the water-holding capacity of farms. Trees also provide natural protection against wind and erosion, enhancing carbon storage while providing shade that reduces temperature and moisture for the benefit of crops or livestock. In regions that

experience heat waves and drought, agroforestry systems are beneficial for farming and climate change.

vii. **Conservation Agriculture**: Conservation farming techniques such as no-till farming, crop rotation, and mulching practices minimize disturbance of the soils, cut evaporation losses, and encourage the rebuilding of the soil structure. These methods increase irrigation water use efficiency and lessen the need for chemicals like fertilizers and pesticides, which makes farming more resilient to climate change.

4.6.2 Strengthening Food Systems

The second approach involves enhancing food systems by locally focusing on improving food production, nutrition access, and the resilience of food supply chain integration. Due to climate changes affecting agricultural yield and food availability, it is pertinent that the following steps be pursued:

1. **Local Food Production**: Encouraging vertical and peri-urban agriculture to enhance food in urban areas is gaining traction. Cities that cultivate food for themselves do not depend on areas set outside their boundaries for meals, which may be affected by climatic instabilities in global food supply chains. Some of Kashif Malthoop's "Gardens in the Sky" and Emmanuel Nyaga's community farming projects in Havana, Bolivia, and Nairobi, Kenya, respectively, have successfully delivered fresh food to local people and cut food mileage and carbon footprints.
2. **Cold Chain Development**: This is necessary to prevent food losses and ensure efficient access to food. Refrigeration units and transport and storage facilities assist in stockpiling perishable food products like fruits, vegetables, dairy, and fish. It is quite helpful for the threatened communities as cold chain systems increase food sufficiency and, more importantly, reduce costs by minimizing post-harvest wastage.
3. **Decentralized Food Processing**: The service provided by small local food processing facilities is very useful in increasing the value of agricultural products, minimizing transportation costs and wastage, and providing employment for the rural area population. For example, small-scale food processing units for drying, milling or canning make it possible to preserve regional economies and guarantee food even in the off-harvest seasons. In India, programs like the National Mission on Food Processing seek to popularize small-scale processing units that improve food availability and increase farmers' incomes.

4.6.3 Mitigation Strategies

4.6.3.1 Reducing Greenhouse Gas Emissions

Adaptation is necessary, but equal emphasis must be placed on measures to reduce greenhouse gas emissions (Filonchyk et al., 2024). Such strategies emphasize the reduction of agriculture, water, and food system emissions while improving carbon sinks to stabilize the Earth's temperature.

Reduction of Greenhouse Gas Emissions in Agriculture: Of all human activities, agriculture occupies one of the largest niches in the emission of carbon gases, accounting for 18–20% of total emissions. Mitigation measures that are key to the reduction of emissions from agriculture include (USEPA, 2024):

a. **Methane Reduction in Livestock**: Cattle, sheep, and goats create a lot of methane through a process known as enteric fermentation (digestion). Cattle farming, in particular, is the most productive. Among the strategies for decreasing methane emissions are improving the quality of livestock feed and using feed additives, like seaweed-based supplements, that can cut methane production by as much as 80%. Additionally, manure management through composting and biogas production can capture methane for use as renewable energy.
b. **Smart Fertilizer Management**: The use of nitrogen fertilizers in excess is a bane because nitrous oxide—a greenhouse gas—gets released. Farmers must learn such farm management practices to apply fertilizers more appropriately and efficiently. With these changes, slow-release fertilizers and nitrification inhibitors can reduce emissions while enhancing crop production.
c. **Rice Farming and Methane Generation**: Methane is emitted to a large extent in the process of flooding the rice fields where rice is grown extensively, e.g., in Southeast Asian regions. The system of Alternate Wetting and Drying (AWD) helps enhance yields while reducing the emission of methane from rice fields. Countries such as Vietnam and the Philippines have successfully adopted AWD in their mitigation frameworks.
d. **Transition to Renewable Energy**: Beyond the coastal areas, the key to reducing emissions in water and food-related systems is the transition from non-renewable to renewable energy sources. Solar-powered irrigation systems (SPIS) are expected to reduce diesel-driven pumps in agriculture and thereby reduce the industry's carbon footprint through improved water use efficiency. Likewise, solar dryers for fruits and biogas production from agricultural waste offer alternative clean energy sources that help reduce over-dependence on fossil fuels.

Small-scale hydropower projects for irrigation or water treatment and management are suitable for plenty of fresh water because they are renewable resources. In addition, these systems can generate electricity that can be utilized in areas such as water irrigation, treatment plants, and rural electrification, thus reducing fossil fuel requirements while enhancing food and water security.

4.6.3.2 Technological Innovations

As for innovative efforts, they could also help with food security in a changing climate via sustainable methods of agriculture and better water management. Advancements in smart water management, growing climate-resilient crops, and developing early warning systems may facilitate better practices and build resilience (Kandel et al., 2024; Saleem et al., 2024).

i. **Smart Water Management Systems**: Through different technologies such as artificial intelligence, the Internet of Things, and remote sensing, water resources can now be monitored and managed effectively and in real time. AI predicts water usage, finds leaks within the systems, and can manage the timing of irrigation systems. IoT devices and sensors also collect water quality utilization data to save water as much as possible.
ii. **Biotechnology for Climate-Resilient Crops**: New possibilities in biotechnology are also making it possible to grow crops that are less sensitive to climate conditions, droughts, heat, or even attacking pests. Genetically modified crops like Bt cotton and drought-tolerant maize are already helping farmers in regions prone to extreme weather conditions, ensuring food security while reducing dependency on water-intensive crops.
iii. **Early Warning Systems**: Early warning systems are very important for preventing damages from extreme climate-related events, including floods, droughts and heat waves. As a result of the improved use of satellite data, meteorological models, and timely monitoring, these can issue warnings in advance, enabling farmers or other people in a community to be prepared for impending danger. When used in areas suffering from food insecurity, these systems can inform the governments about the estimated poor harvest and prepare food assistance and strategies to sharpen the relief efforts.

4.7 Challenges and Barriers to Addressing Climate Change, Water, and Food Security

While there are several relevant adaptation and mitigation approaches to cope with the impacts of climate change on water and food security, a number of challenges and barriers exist that score progress (Seppelt et al., 2022). These barriers are complex and include socioeconomic, policy, and knowledge related to barriers that need to be resolved to achieve sustainable development and systems resilience to climate change in the future.

4.7.1 Socioeconomic Barriers

i. **Poverty and Inequality**: Climatic changes and subsequent water and food insecurity are caused by poverty and remain one of the main deterrents. Poor, vulnerable sections of the population, especially in low-income and rural areas, do not have the financial resources to spend on climate resilience construction, including the creation of effective irrigation structures, water carts or better seeds. Without sufficient access to these resources, vulnerabilities to climate change effects or impacts are increased with droughts, floods, and extreme temperatures, particularly affecting agricultural and water availability.
Specifically, small farmers in the sub-Saharan region of Africa depend on rainfall for their farming, which is very much prone to climate variability. Such farmers without irrigation facilities or drought-resistant crops suffer from crop losses, low income and inadequate food. The climate impact traps them within the poverty trap since climate change impact further shrinks their scope to recover from or adapt to the impacts (Omotoso et al., 2023).

ii. **Lack of Infrastructure**: Infrastructure in the water and food systems remains a common challenge in many areas. Such poor water campaigning infrastructure includes ineffective irrigation, poor facilities for water storage, or even distribution systems that bias people's efforts to improve efficiency on water use. Farmers with few nodes of contact to the markets, transport systems, and cold chain storage facilities in the countryside and rural areas make them vulnerable to food insecurity. On the other hand, post-harvest losses are drastically high in regions with a lack of appropriate infrastructure meant for combating such losses, impeding food security and wastage efficiencies (Mirzabaev et al., 2023).
In places where water is in short supply, people often turn to already depleted aquifers, and where no efficient recharge or management techniques are employed, aquifers may eventually run dry. On the same note, food transport and storage gaps mean that some harvested crops may deteriorate before being put to use or sold, thus escalating food wastage and economic impacts.

iii. **Social Vulnerabilities in the Context of Gender and Climate Change**: Gender discrimination increases the negative effects of climate change, especially in disadvantaged communities and rural regions. Water collecting and food production in the household—mainly the responsibility of women in developing countries—becomes increasingly burdensome. Keeping women away from productive resources like land, credit, and education or excluding them from decision-making frequently leaves them no choice but to skip the more sensitive climate-smart technologies and investment in water management techniques.
For example, in areas of water shortages, it becomes a norm that women and children transverse extensive distances to fetch water, which can be time-consuming. This also makes it impossible for them to perform other productive roles like

agriculture, schooling, or income-generating activities. It is important to highlight that women must be empowered with resources, education, and decision-making opportunities to improve climate resilience and ensure achieving food and water security for all.

4.7.2 Policy and Governance Barriers

i. **Lack of Coordination and Integrated Policy**: The fragmented nature of the policy-making process is an overarching barrier to tackling water and food security in the context of climate change. While water, food, and energy systems are related, policies tend to address one issue at a time, creating inefficiencies and gaps for integrated solutions. For instance, water management policies may cater more to industrial or urban water usage while the water-intensive agriculture sectors' requirements are ignored. Similarly, agricultural policies may not emphasize adopting water-efficient irrigation systems or neglect to factor in water shortages in food supply strategies.
Consequentially, it is important to shift from managing water, food and energy resources separately to managing them as a single entity. In this context, the Water-Energy-Food Nexus framework comprehensively posits and understands such challenges while supporting policies integrating these sectors. Hence, governments must develop intersectoral policies that harmonize water management with agricultural and energy requirements to ensure that food and water resources are secure under predicting climate changes.

ii. **Political Will and Governance Challenges**: There is sometimes a lack of political will in aiding water and food security from the climate change perspective in long-term strategies. In short, political election "cycles", priority conflicts, and funding deficits could slow down or disallow urgent policies and actions. In some places, even neglecting climate change or denying it in the first place, the greatest threat to progress is the political leadership, who opt for short-term thinking in economic development. In the end, this results in poor use of resources, high emissions of greenhouse gases, and, in some cases, excessive and wasteful agricultural practices that threaten food and water security.
Also, corruption and poor implementation due to ineffective governance systems may undermine certain policies. In certain nations, there's chronic mismanagement and over-extraction of water and its associated resources because of inappropriate allocations, which tend to disadvantage the marginalized and increase water stress on the already disadvantaged segments. Good governance, accountability, and transparency must be established to ensure that policies are properly devised and executed and resources are apportioned as intended (Lee et al., 2022).

iii. **Lack of Global Collaboration**: Combating climate change and global warming is not only a topic for each country to address within its borders. Developing countries, however, face many challenges in negotiating and implementing

universal treaties. Resources and access to technology in developing countries present clear challenges to countries achieving global targets such as the Paris Accord. Additionally, transboundary water resources often lead to conflicts between countries, as seen with the Nile, Mekong, and Indus rivers.
International cooperation is essential to address these challenges. Shared water resources must be managed cooperatively to avoid conflicts and ensure equitable access. Global efforts to support developing countries with technology, finance, and capacity building will enable them to address water and food security challenges while mitigating climate change (UNDP, 2024).

4.7.3 *Knowledge Gaps and Resistance to New Practices*

i. **Information Gaps and Research Deficiencies:**
Notably, many developing countries need adequate data on water availability, food security levels, and climate change effects, as such data is absent. Without appropriate and relevant information, government officials, non-governmental organizations and even communities cannot formulate appropriate strategies and policies. The absence of climate data is more pronounced in the areas that are already more affected and are bound to be more affected in the future due to climate change. Such areas require information on the availability of rainfall, water resources, and appropriate decisions regarding what crop to grow.
Further, a need to conduct more localized studies on climate-resilient agriculture and sustainable water-managed practices arises. The impact of climate change, for instance, is not uniform across the world; thus, context-specific approaches must be employed. Unfortunately, research funding for these localized solutions has been so poor that it has hindered the development and application of effective measures to adapt and mitigate its effects.
ii. **Resistance to the Adoption of New Practices**: At times, there are barriers when it comes to deploying new techniques in agriculture and water management and climate adaptation measures, and more often, this is pronounced in rural areas where there are unapologetically rigid customs. Some farmers are also reluctant to use improved technologies, including drought-resistant seed varieties and modern irrigation methods, due to lack of exposure, characteristics of high perceived costs, and even culture (Wakweya, 2023).
For example, certain farmers may wish to adopt drip irrigation systems, although they may not be familiar with the technology, or the initial investment may not be feasible. A similar case can be made for climate-resilient crops, which include genetically modified and drought-tolerant varieties, which may be limited due to various safety, cost, or perceived benefit issues. These challenges can be overcome by education, training and other forms of support so that farmers appreciate the long-term benefits of such practices and technologies.

4.8 Education and Capacity Building

Building resilience to climate change requires education and capacity-building investment efforts at all levels. Providing farmers, water managers, and local communities with the capacity to adopt any climate adaptation strategies is important in augmenting the effectiveness of the mitigation measures. Communities can be further educated through training on water-saving irrigation, climate-positive farming, and sustainable land use, which will improve their capacity to deal with the effects of climate change.

For example, training farmers in dry regions, such as the Sahel in Africa, on using water-efficient irrigation and climate-friendly agricultural practices has helped increase yields and minimize the risks of droughts. Protecting indigenous knowledge and giving people the resources they require for evolving change is integral to the fight against climate change.

4.9 Future Directions and Research Prospectives

As the world grapples with climate change, water insecurity, food security, and other future directions, dealing with these aligned issues must be interdisciplinary. Research, policy formulation and local community activities should be congruent to enhance resilience, sustainability and food and water security.

i. **Need for Integrated Approaches**: Future research should focus on the links between water, food, and climate policies. The Water-Energy-Food Nexus paradigm offers a more integrated solution by addressing the inter-relationships among these sectors. Such policies should be designed in the future to sustainably relate the three resources of water, food, and energy.
   Multilateral policies are needed to bring together governments, NGOs, and private entities and consider the connections between water, food, and climate. We can address these systems in a more efficient, less conflictual, and fairer way to all concerned if we manage them in an integrated way.
ii. **Multi-stakeholder Collaboration**: Climate change affects everyone, so it must be addressed cooperatively. The challenge of water and food security in climate change calls for all stakeholders, including government, private sector, NGOs, and local people. Strong collaboration, particularly between the state and business, forces the resources and experience necessary to expand the scope of effective strategies and help vulnerable groups.
   In addition, local people need to be empowered to take responsibility for water and food security issues. Participatory governance and the inclusion of indigenous knowledge systems can enhance the sustainability of water and food systems, ensuring that solutions are context-specific and locally driven.

Emerging Threats: Exploring new challenges, such as disasters and diseases caused by climate change, environmental migration, and loss of biodiversity, is of utmost importance for future planning and strategies for adaptation. The advent of new pests and diseases, reduction in the number of important pollinators and movement of populations due to climate change will all pose continuous threats to water and food security in the decades to come. Developing crops that can withstand climate shocks, water-conservative technologies, and best resource management practices will help address and build resilience to such risks.

Capacity Building: Education and capacity building are the other effective ways to enhance adaptive capacity to climate change. It is also important that farmers, communities, and governments have the instruments, knowledge, and resources to execute climate-smart strategies and ensure sustainability in the long run.

4.10 Conclusion

The complex and interdependent relationship between climate change, water, and food security poses one of the major critical threats at present. The earth's temperature is continuously rising, leading to the change in weather patterns and precipitation intensity. It affects the earth's ecosystems, which determine the livelihoods of the people directly and indirectly dependent on natural climatic conditions for crop cultivation and production. This chapter highlighted the impact of climate change on food production and the declining availability of water resources to grow and cultivate crops. The declining water resources lead to a decline in agricultural productivity, which affects the supply chains and, ultimately, the economy of the region.

The obstacles are scary but not formidable. Sustainable agriculture and aquaculture practices like climate-smart and precision farming and efficient utilization of renewable resources can prove very promising in the long run. However, implementing these approaches requires strong cooperation and interdependence between climate change countering authorities, water management, and food production organizations at regional and global levels.

In developing continents like Asia, Africa, and Small Island Developing States, a higher resistance toward adopting better management practices remains a significant challenge for international organizations tackling climate change, water resources management and food scarcity. There is a need to bridge the gap between adopting these practices by involving multiple stakeholders, empowering marginalized communities, developing capacity-building programs, women empowerment, and decision-making participation to manage natural resources sustainably.

Future research must efficiently utilize AI-driven technological advancement that aligns with climate-resilient practices. Cooperation between international organizations and good governance at the local and regional level is a step that needs to be taken to achieve the Sustainable Development Goals (SDGs), particularly those related to water and food security and climate change. It will ensure the judicious use of these resources, safeguard the food production systems and water resources, and mitigate the risks related to climate change, which are the fundamental resources and rights of future generations.

References

Abbass, K., Qasim, M. Z., Song, H., Murshed, M., Mahmood, H., & Younis, I. (2022). A review of the global climate change impacts, adaptation, and sustainable mitigation measures. *Environmental Science and Pollution Research, 29*(28), 42539–42559. https://doi.org/10.1007/s11356-022-19718-6

Adjah, K. L., Asante, M. D., Toure, A., Aziadekey, M., Amoako-Andoh, F. O., Frei, M., Diallo, Y., & Agboka, K. (2022). Improvement of rice production under drought conditions in West Africa: Application of QTLs in breeding for drought resistance. *Rice Science, 29*(6), 512–521. https://doi.org/10.1016/j.rsci.2022.06.002

Baocheng, H., Jamil, A., Bellaoulah, M., Mukhtar, A., & Clauvis, N. K. (2024). Impact of climate change on water scarcity in Pakistan. Implications for water management and policy. *Journal of Water and Climate Change, 15*(8), 3602–3623. https://doi.org/10.2166/wcc.2024.710

Bhattarai, N., Lobell, D. B., Balwinder-Singh, Fishman, R., Kustas, W. P., Pokhrel, Y., & Jain, M. (2023). Warming temperatures exacerbate groundwater depletion rates in India. *Science Advances, 9*(35), 1–18. https://doi.org/10.1126/sciadv.adi1401

Bouteska, A., Sharif, T., Bhuiyan, F., & Abedin, M. Z. (2024). Impacts of the changing climate on agricultural productivity and food security: Evidence from Ethiopia. *Journal of Cleaner Production, 449*, 1–19. https://doi.org/10.1016/j.jclepro.2024.141793

Cheung, W. W., Frölicher, T. L., Lam, V. W., Oyinlola, M. A., Reygondeau, G., Sumaila, U. R., Tai, T. C., Teh, L. C., & Wabnitz, C. C. (2021). Marine high temperature extremes amplify the impacts of climate change on fish and fisheries. *Science Advances, 7*(40), 1–20. https://doi.org/10.1126/sciadv.abh0895

Dao, P. U., Heuzard, A. G., Le, T. X. H., Zhao, J., Yin, R., Shang, C., & Fan, C. (2024). The impacts of climate change on groundwater quality: A review. *Science of the Total Environment, 912*, 1–15. https://doi.org/10.1016/j.scitotenv.2023.169241

Dar, M. H., Waza, S. A., Shukla, S., Zaidi, N. W., Nayak, S., Hossain, M., Kumar, A., Ismail, A. M., & Singh, U. S. (2020). Drought tolerant rice for ensuring food security in Eastern India. *Sustainability, 12*(6), 2214. https://doi.org/10.3390/su12062214

Ebi, K. L., Vanos, J., Baldwin, J. W., Bell, J. E., Hondula, D. M., Errett, N. A., Hayes, K., Reid, C. E., Saha, S., Spector, J., & Berry, P. (2021). Extreme weather and climate change: Population health and health system implications. *Annual Review of Public Health, 42*(1), 293–315. https://doi.org/10.1146/annurev-publhealth-012420-105026

FAO. (2021). The state of the world's land and water resources for food and agriculture Systems at breaking point. *Synthesis report 2021*.https://doi.org/10.4060/cb9910en

Filonchyk, M., Peterson, M. P., Yan, H., Gusev, A., Zhang, L., He, Y., & Yang, S. (2024). Greenhouse gas emissions and reduction strategies for the world's largest greenhouse gas emitters. *Science of The Total Environment*, 1–18. https://doi.org/10.1016/j.scitotenv.2024.173895

Fisher, M., Abate, T., Lunduka, R. W., Asnake, W., Alemayehu, Y., & Madulu, R. B. (2015). Drought tolerant maize for farmer adaptation to drought in sub-Saharan Africa: Determinants of adoption in eastern and southern Africa. *Climatic Change, 133*, 283–299. https://doi.org/10.1007/s10584-015-1459-2

Gul, S., Uzair, S., Shakir, A. M., Gowhar, I., & Hazi, Q. L. (2024). Enhancing global food security through sustainable fisheries and aquaculture: A comprehensive review. *Asian Journal of Agricultural Extension, Economics & Sociology, 42*(10), 60–70. https://doi.org/10.9734/ajaees/2024/v42i102563

Han, J. W., Zuo, M., Zhu, W. Y., Zuo, J. H., Lü, E. L., & Yang, X. T. (2021). A comprehensive review of cold chain logistics for fresh agricultural products: Current status, challenges, and future trends. *Trends in Food Science & Technology, 109*, 536–551. https://doi.org/10.1016/j.tifs.2021.01.066

Hoegh, O., Jacob, D., Taylor, M., Bindi, M., Brown, S., Camilloni, I. A., Diedhiou, A., Djalante, R., Ebi, K. L., Engelbrecht, F., & Guiot, J. (2018). *Impacts of 1.5 C global warming on natural and human systems* (pp. 175–312). Cambridge University Press. https://doi.org/10.1017/9781009157940.005

IPCC. (2021). Intergovernmental Panel on Climate Change. Retrieved December 25, 2024, from https://www.ipcc.ch/assessment-report/ar6/

Kandel, G. P., Bavorova, M., Ullah, A., & Pradhan, P. (2024). Food security and sustainability through adaptation to climate change: Lessons learned from Nepal. *International Journal of Disaster Risk Reduction, 101*, 1–16. https://doi.org/10.1016/j.ijdrr.2024.104279

Kitonga, K., Jamora, N., Smale, M., & Muchugi, A. (2020). Use and benefits of tree germplasm from the World Agroforestry genebank for smallholder farmers in Kenya. *Food Security, 12*(5), 993–1003. https://doi.org/10.1007/s12571-020-01047-6

Lee, S., Paavola, J., & Dessai, S. (2022). Towards a deeper understanding of barriers to national climate change adaptation policy: A systematic review. *Climate Risk Management, 35*, 1–18. https://doi.org/10.1016/j.crm.2022.100414

Makule, E., Dimoso, N., & Tassou, S. A. (2022). Precooling and cold storage methods for fruits and vegetables in Sub-Saharan Africa—A review. *Horticulturae, 8*(9), 776. https://doi.org/10.3390/horticulturae8090776

Mirzabaev, A., Kerr, R. B., Hasegawa, T., Pradhan, P., Wreford, A., von der Pahlen, M. C. T., & Gurney-Smith, H. (2023). Severe climate change risks to food security and nutrition. *Climate Risk Management, 39*, 1–15. https://doi.org/10.1016/j.crm.2022.100473

Mustafa, M. F. M. S., Namasivayam, N., & Demirovic, A. (2024). Food cold chain logistics and management: A review of current development and emerging trends. *Journal of Agriculture and Food Research*, 101343. https://doi.org/10.1016/j.jafr.2024.101343

Nyaga, J., Muthuri, C. W., Barrios, E., Öborn, I., & Sinclair, F. L. (2019). Enhancing maize productivity in agroforestry systems through managing competition: Lessons from smallholders' farms, Rift valley, Kenya. *Agroforestry Systems, 93*, 715–730. https://doi.org/10.1007/s10457-017-0169-3

Omotoso, A. B., Letsoalo, S., Olagunju, K. O., Tshwene, C. S., & Omotayo, A. O. (2023). Climate change and variability in sub-Saharan Africa: A systematic review of trends and impacts on agriculture. *Journal of Cleaner Production, 414*, 1–14. https://doi.org/10.1016/j.jclepro.2023.137487

Pizzorni, M., Innocenti, A., & Tollin, N. (2024). Droughts and floods in a changing climate and implications for multi-hazard urban planning: A review. *City and Environment Interactions*, 100169. https://doi.org/10.1016/j.cacint.2024.100169

Puppala, H., Peddinti, P. R., Kim, B., & Arora, M. K. (2023). Unmanned aerial vehicles for planning rooftop rainwater harvesting systems: A case study from Gurgaon, India. *Water Supply, 23*(5), 2014–2030. https://doi.org/10.2166/ws.2023.105

Rois-Diaz, M., Lovric, N., Lovric, M., Ferreiro-Domínguez, N., Mosquera-Losada, M. R., Den Herder, M., Graves, A., Palma, J. H., Paulo, J. A., Pisanelli, A., & Smith, J. (2018). Farmers' reasoning behind the uptake of agroforestry practices: Evidence from multiple case-studies across Europe. *Agroforestry Systems, 92*, 811–828. https://doi.org/10.1007/s10457-017-0139-9

Saleem, A., Anwar, S., Nawaz, T., Fahad, S., Saud, S., Ur Rahman, T., Khan, M. N. R., & Nawaz, T. (2024). Securing a sustainable future: The climate change threat to agriculture, food security, and sustainable development goals. *Journal of Umm Al-Qura University for Applied Sciences*, 1–17. https://doi.org/10.1007/s43994-024-00177-3

Seppelt, R., Klotz, S., Peiter, E., & Volk, M. (2022). Agriculture and food security under a changing climate: An underestimated challenge. *Iscience, 25*(12). https://doi.org/10.1016/j.isci.2022.105551

Shah, T., & Lele, U. (2011). Climate change, food and water security in South Asia: Critical issues and cooperative strategies in an age of increased risk and uncertainty. In *Synthesis of workshop discussions. A Global Water Partnership (GWP) and International Water Management Institute (IWMI) workshop* (pp. 23–25). https://www.gwp.org/globalassets/global/about-gwp/publications/colombo-synthesis-report-climate-change-food-and-water-security-in-south-asia-final.pdf

Sharma, U., Patwardhan, A., & Parthasarathy, D. (2009). Assessing adaptive capacity to tropical cyclones in the East coast of India: A pilot study of public response to cyclone warning information. *Climatic Change, 94*(1), 189–209. https://doi.org/10.1007/s10584-009-9552-z

Smith, M. D., Sikka, A., Dirwai, T. L., & Mabhaudhi, T. (2023). Research and innovation in agricultural water management for a water-secure world. *Irrigation and Drainage, 72*(5), 1245–1259. https://doi.org/10.1002/ird.2872

Tran, T. L., Ritchie, E. A., & Perkins-Kirkpatrick, S. E. (2022). A 50-year tropical cyclone exposure climatology in southeast Asia. *Journal of Geophysical Research: Atmospheres, 127*(4), 1–40. p.e2021JD036301. https://doi.org/10.1029/2021JD036301

UN DESA. (2024). United Nations Department of Economic and Social Affairs. World population projected to reach 9.8 billion in 2050, and 11.2 billion in 2100. Retrieved December 30, 2024, from https://www.un.org/en/desa/world-population-projected-reach-98-billion-2050-and-112-billion-2100

UNDP. (2024). United Nations Development Programme. Global cooperation on climate change and sustainable development. Retrieved December 30, 2024, from https://climate-box.com/textbooks/3-how-to-prevent-dangerous-climate-change/3-5-global-cooperation-on-climate-change-and-sustainable-development/

UNEP. (2023). United Nations Environment Programme. Emissions Gap Report 2023: Broken Record—Temperatures hit new highs, yet world fails to cut emissions (again). Nairobi. https://doi.org/10.59117/20.500.11822/43922

UNEP. (2024). United Nations Environment Programme. Emissions Gap Report 2024: Climate Change Action Reports. https://doi.org/10.59117/20.500.11822/46404

UNEPA. (2024). United Nations Environmental Protection Agency. Retrieved November 20, 2024, from https://www.epa.gov/climate-indicators/weather-climate

USEPA. (2024). United States Environmental Protection Agency. Sources of Greenhouse Gas Emissions. Retrieved November 20, 2024, from https://www.epa.gov/ghgemissions/sources-greenhouse-gas-emissions

Vousdoukas, M. I., Athanasiou, P., Giardino, A., Mentaschi, L., Stocchino, A., Kopp, R. E., Menéndez, P., Beck, M. W., Ranasinghe, R., & Feyen, L. (2023). Small island developing states under threat by rising seas even in a 1.5° C warming world. *Nature Sustainability, 6*(12), 1552–1564. https://doi.org/10.1038/s41893-023-01230-5

Wakweya, R. B. (2023). Challenges and prospects of adopting climate-smart agricultural practices and technologies: Implications for food security. *Journal of Agriculture and Food Research*, 1–8. https://doi.org/10.1016/j.jafr.2023.100698

Xu, R., Yu, P., Liu, Y., Chen, G., Yang, Z., Zhang, Y., Wu, Y., Beggs, P. J., Zhang, Y., Boocock, J., & Ji, F. (2023). Climate change, environmental extremes, and human health in Australia: Challenges, adaptation strategies, and policy gaps. *The Lancet Regional Health-Western Pacific, 40*, 1–16. https://doi.org/10.1016/j.lanwpc.2023.100936

Zhou, H., Tao, F., Chen, Y., Yin, L., Wang, Y., Li, Y., & Zhang, S. (2024). Climate change reduces agricultural total factor productivity in major agricultural production areas of China even with continuously increasing agricultural inputs. *Agricultural and Forest Meteorology, 349*, 1–17. https://doi.org/10.1016/j.crm.2022.100473

Chapter 5
Global Food Systems—The Challenges and Trends in the Anthropocene Epoch

Deepak Beniwal, Vibhuti Gulia, Sahil Singh Dhull, and Jyoti Rani

Abstract The epoch, therefore, presents a mix of challenges and opportunities for the global food system. In contrast, anthropogenic scale changes in food and agriculture are critical to understanding the Anthropocene as both a sequence of physiological events and an existential crisis of contemporary society. Food systems have the potential to enhance human health and support environmental sustainability; nevertheless, they now undermine both. Meeting the nutritional needs of an expanding global population through sustainable food systems is urgent. Agri-food systems play an important role in addressing serious environmental trends and may directly confront significant challenges associated with the Anthropocene concept. This chapter examines the essential themes that characterize the interplay between people and the environment, emphasizing the influence of food production, consumption, and distribution on climate change, biodiversity decline, and the depletion of natural resources. Primary challenges encompass unsustainable farming methods, food insecurity, waste management, and increasing population pressures. This chapter also explores upcoming trends including sustainable agriculture, technological innovation, plant-based diets, and circular economy models. It concludes by examining possible strategies for reforming the food system to reduce environmental damage while maintaining food security amid rapid ecological change.

Keywords Anthropocene epoch · Food system · Sustainable agriculture · Climate change · Food insecurity

---

D. Beniwal · V. Gulia · S. S. Dhull · J. Rani (✉)
Department of Botany, Chaudhary Devi Lal University, Sirsa 125055, India
e-mail: drjyotisangwan@cdlu.ac.in

D. Beniwal
e-mail: seenubnwl03@gmail.com

V. Gulia
e-mail: vibs.vibhuti@gmail.com

S. S. Dhull
e-mail: sahil75dhull@gmail.com

J. Parray and A. K. Haghi (eds.), *Food and Water Security*,
https://doi.org/10.1007/978-3-032-04870-7_5

5.1 Introduction

Human-induced alterations to the land, seas, Earth's climate, and biosphere have become so significant and swift that they warrant the designation of a brand-new geological period, termed the Anthropocene. In the past decade, the notion of the Anthropocene has gained significant recognition inside and beyond geoscientific literature, however, its parameters remain ambiguous (Gibbard et al., 2022). The proposition for the "Anthropocene" period as an official unit of the geologic time scale has garnered significant interest in both scientific and public discourse. Nevertheless, several writings regarding the Anthropocene inaccurately depict the characteristics of the units within the International Chronostratigraphic Chart, which is generated by the International Commission on Stratigraphy (ICS) and is the foundation for the geologic time scale. The stratigraphic record of the Anthropocene is limited, particularly given its recently suggested commencement in 1945; it corresponds to a human lifespan, and this definition consigns significant anthropogenic alterations to a "pre-Anthropocene" (Finney & Edwards 2016).

Indeed, human endeavors leave their ecological footprints and modify the world's primary biogeochemical cycles as well as the course of evolution. Atmospheric concentrations of carbon dioxide (CO_2), which stand well above 280 parts per million (ppm) in the atmosphere today, have reached levels higher than the threshold for the last 800,000 years, because of anthropogenic activities. Since 1750, human activities have accounted for the emission of 555 petagrams of carbon to the atmosphere thus, delaying the next glacial event to occur on planet Earth (Lewis & Maslin 2015). In many areas, the pressure of man became increasingly evident during the closing stages of the Pleistocene era when linked with the mass death of large animals; the loss of the saber-toothed tiger in North America and of the woolly mammoths in Siberia are among the common examples. The extinction of megafauna on several continents seems to align with the emergence of modern humans. This extinction is diachronous, occurring at many locations and dates, similar to numerous occurrences in the geological record. The megafauna vanished in Australia 50,000 years ago, but in the Americas, it occurred 13,000 years ago. Nevertheless, megafauna continue to inhabit some regions of Africa and South Asia, although they face threats in practically all locations. After the Neolithic agricultural revolution, humans began living in villages and towns, and by the third millennium BC, the cities of ancient Mesopotamia, the Nile Valley, and the Indus Basin of Pakistan were firmly established and culturally unique. After that, urban civilizations spread in all tropical and temperate regions of the world; those in Europe, Central and South America, and China became diverse and complex by the first millennium BC. The pace of urbanization accelerated over time, and the first megacity with a population of a million could have been formed as early as in the late medieval period. By the eighteenth century, Paris and London had attained this magnitude. Currently, several cities possess populations ranging from 10 to 20 million residents. These are seeing significant growth. The actual consequence of urbanization has been a natural impact that results from increased populations. From 1800 the world population swelled up to 1 billion while in 2000 stands at 6.5 billion and should it persist, it may have crossed 9 billion people in 2050 (Zalasiewicz et al., 2011).

Due to the widespread presence of agriculture, it is probably not unexpected that the environmental effects of this industry are regularly cited as proof that a new geologic epoch—the "human dominated geological epoch" known as the Anthropocene—has begun. Even though a large and expanding literature of research has highlighted the crucial role that agriculture plays in global environmental change, ranging from soil to water to atmospheric carbon, it has not always engaged with criticisms of the Anthropocene theory. One of the most well-known academic attempts to discuss agri-food systems in this new assumed era is the EAT-Lancet Commission report on Food in the Anthropocene, which takes the term "Anthropocene" at face value and suggests a "universal healthy reference diet" to lessen the effects of a time when "humanity" has taken the lead in bringing about change (Reisman & Fairbairn, 2020). The Anthropocene epoch is associated with worldwide food hazards, like, antibiotic-resistant pathogens, mass breakouts of foodborne illnesses, and bovine spongiform encephalopathy. The eating habits and the human-centered meat industry have added new food security risks and also increased the frequency of previously occurring demands (Lougheed & Hird, 2017). Hird and Yusoff believe that the Anthropocene crisis is mainly a "crisis of waste." This involves the unwanted products of modern life, such as the loss of plants, fossil fuels, water, animal species, air pollution, and waste from things like shells, land, skin, plants, and tiny organisms. It also includes waste from efforts to protect our food supply, which often require even more protection measures. Keeping the food system safe usually means trying—and often failing—to manage human behavior and the growing number of bacteria, viruses, prions, machines, and other parts of the system. Materials like meat, plants, packaging waste, liquid from landfills, tiny organisms, and antibiotics can pose serious risks to our security and safety (Hird & Yusoff, 2016).

If providing food and nutritional security to everyone over an extended period with little ecological harm is the aim of a sustainable food system, then the biophysical foundation for revolutionary transformation is clear. Roughly one-third of Earth's surface is covered by crops, and 26% of the emissions of greenhouse gases worldwide are caused by agricultural activities. Food systems are the biggest source of freshwater consumption, the main cause of ecological degradation and biodiversity loss, and significant sources of several contaminants, antibiotics, such as microplastics, nitrogen, phosphorus, and heavy metals (Grumbine et al., 2021). With an eye toward the future of food in the Anthropocene, this chapter aims to provide a comprehensive overview of the issues and trends while delving into the linkages. Creative approaches to sustainable practices entail moving forward with plans for an even more substantial food system that might meet the demands of the current crises while striking a suitable balance between the requirements of people and the environment. In a world undergoing unprecedented change, we must reconsider how food should be produced, shared, and consumed since the decisions we make now will influence the foodscapes of tomorrow.

5.2 Challenges in the Food System

Across the globe, various challenges have obstructed the efficient supply of food systems for a growing population in the future. Climate change, resource scarcity, and economic disparities represent some of the numerous challenges impacting food production and distribution. These issues are interconnected, leading to a range of challenges that necessitate innovative and holistic strategies for resolution. It is important to understand that these issues are closely related to equity in society, environmental preservation, and enhancing food access for all societies. Some of the main challenges to the food system are represented in Fig. 5.1.

5.2.1 Unsustainable Agriculture Practices

Modern agriculture commonly uses high-input farming technologies with extremely huge pressure on natural resources. Such strategies strip critical nutrients from the soil, exhaust ground and surface waters, and rely heavily on chemical inputs, like synthetic fertilizers and pesticides. This eventually results in high levels of erosion of soil, pollution of water bodies, and a decrease in the long-term productivity of

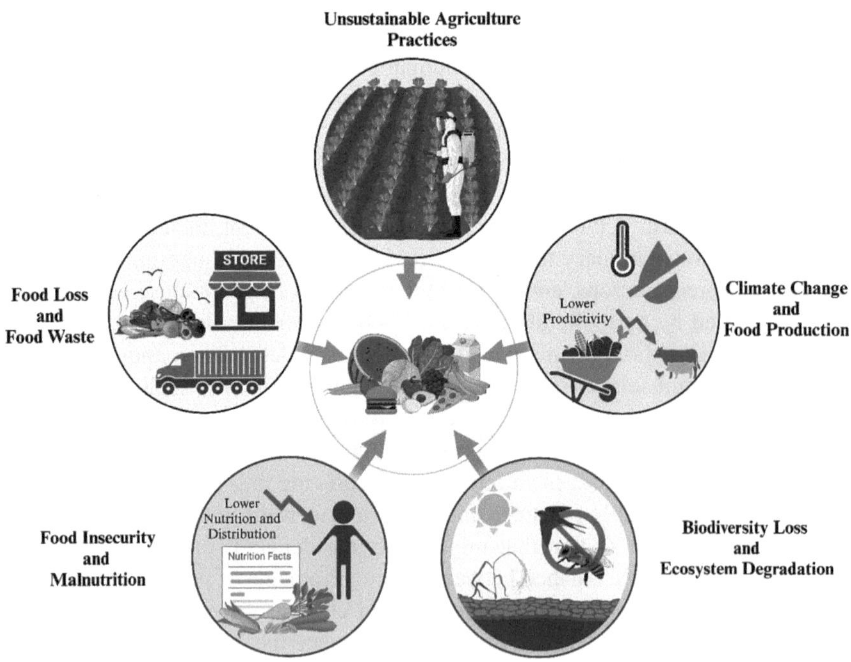

Fig. 5.1 Challenges in food system

agriculture. Major environmental risk factors here include the high application of synthetic chemicals, particularly fertilizers, and pesticides. For instance, the excess application of fertilizer has risks such as water pollution, while pest-resistant pests and herbicide-resistant weeds are some other products of similar excesses (Hossain et al., 2020). Monoculture farming that involves the mass cultivation of one crop in multiple areas reduces biodiversity. Biodiversity is important for pest control, soil health, and ecosystem stability. As a result, crop loss from genetic diversity makes it susceptible to diseases and pests. This is an enormous risk to the country's vast failure in agriculture. The additive effect of these unsustainable systems threatens the very base of food production worldwide because the degradation of natural systems undermines the potential for sustainable agricultural input (Mrabet, 2023).

5.2.2 *Climate Change and Food Production*

Climate change, appearing as increased global temperatures, altered precipitation regimes, and intensified extreme events, including heatwaves, droughts, and floods, is particularly vulnerable to the agricultural sector. These climatic fluctuations introduce some level of uncertainty in food production that creates unrest for farmers in preparing planting and harvesting timetables. The weather may also shorten the growing period of some crops and exacerbate stressors in others because crop pests and diseases are more severe in the warm season (Gomez-Zavaglia et al., 2020). Droughts can devastate crop yields either because they become very frequent or because of their severity, and flooding can achieve the same effect because fertile topsoil is washed away and planting cycle is disrupted. Livestock productivity may also decline due to heat stress as feed and water become less readily available. This volatility in agricultural output has very tremendous implications for food security as lesser yields lead to shortages, increased prices, and a break in the food chain, especially in unstable regions economically or politically (Fresco, 2009).

5.2.3 *Biodiversity Loss and Ecosystem Degradation*

One of the most important indirect effects of industrial agriculture is habitat destruction and loss of biodiversity from land clearing for crop and livestock production. Expansion leads to land for agriculture, resulting in deforestation and changing diverse ecosystems into monocultures or degraded landscapes. Such reduction in biodiversity might help a few species but directly affects agriculture negatively. Natural predators and pollinators like bees and birds are examples where loss can instantly devastate crops for higher yields (Scherer et al., 2020). In addition, degraded ecosystems are less resilient to environmental changes, such as climate shifts or invasive species. Disturbances in natural food webs and ecological functions can produce an effect of ripples as it gives an agricultural system ample time and opportunity to be more vulnerable to pests and diseases and to further challenges. Ultimately, this degradation impairs the sustainability of the systems of food production and threatens global food stability (Gupta, 2019).

5.2.4 Food Insecurity and Malnutrition

Despite technological advancement and an increase in food production globally, millions continue to suffer from hunger and malnutrition. This is not a problem of low production but rather of access and equal distribution. Economic inequality, poverty, conflict, and inadequate infrastructure stop the majority of people in this world from getting adequate, not to mention nutritional food. Poor access to productive resources in developing countries combined with poor agricultural practices that do not favor productivity leads to chronic hunger accompanied by micronutrient deficiencies among many communities (Von Braun, 2010). Those trigger several health implications among children, including stunted growth, and increased vulnerability to diseases. While it is also common in affluent settings, where a consumption pattern of processed and ultra-processed foods has contributed to growing levels of obesity and associated problems, it generates paradoxical conditions wherein undernutrition and overnutrition coexist. Thus, the "double burden" of malnutrition poses a challenge of a different type for public health systems, again stressing the complex interrelation between food supply, dietary practices, and health outcomes (Alaimo et al., 2020).

5.2.5 Food Loss and Waste

The most outstanding threats in the global food system have been the high losses and waste at different levels down the supply chain from production to consumption. More than in the developed countries, food wastage in developing countries is ascribed to the lack of infrastructures adequately sufficient with low storage facilities, ill-conceived transport networks, and inappropriate handling practices that result in a large percentage of wastage even before they reach the markets (Thomas, 2022). In developed countries, food wastage tends to occur more at the retail and consumer levels because of factors such as overproduction, cosmetic standards for fresh fruits and vegetables, and consumer behavior say edible food wasted. Environmental degradation is significant because the water, energy, and land used non-productively, add to the greenhouse gas emissions. Most of this decomposes in landfills, producing methane, a highly potent greenhouse gas. At all points of the food supply chain, losses and waste need to be minimized in order to build up improved food security and diminish negative impacts from agriculture and on food (Mazac & Tuomisto, 2020).

5.3 Emerging Trends in the Food System

Technological developments and changing consumer habits are causing significant changes in the global food chain. The trends highlighted include a focus on alternative proteins, the advancement of customized nutrition, and the growing accessibility of food safety and quality, as illustrated in Fig. 5.2. The entire food supply chain, from production to consumption, is being influenced by these trends. The management of food is increasingly efficient and less wasteful, driven by advancements in technology, together with the emergence of new marketing channels for food purchases and consuming food. The influence of these evolving trends will likely result in a more balanced, efficient, and adaptable food system.

5.3.1 *Sustainable Agricultural Innovations*

For long enough, agricultural sectors have been convicted of aggravating climate change and simultaneously being victimized by the same effects. It accounts for the largest portion of the world's emissions of greenhouse gas. However, the sector is not free from the same catastrophic effects it has caused-including, but not limited to, changes in the weather, land degradation, and water scarcity. This is hence where sustainable agricultural innovations come at the heart of the fight for a more sustainable and environmentally friendly food system (Rockström et al., 2016). One of the

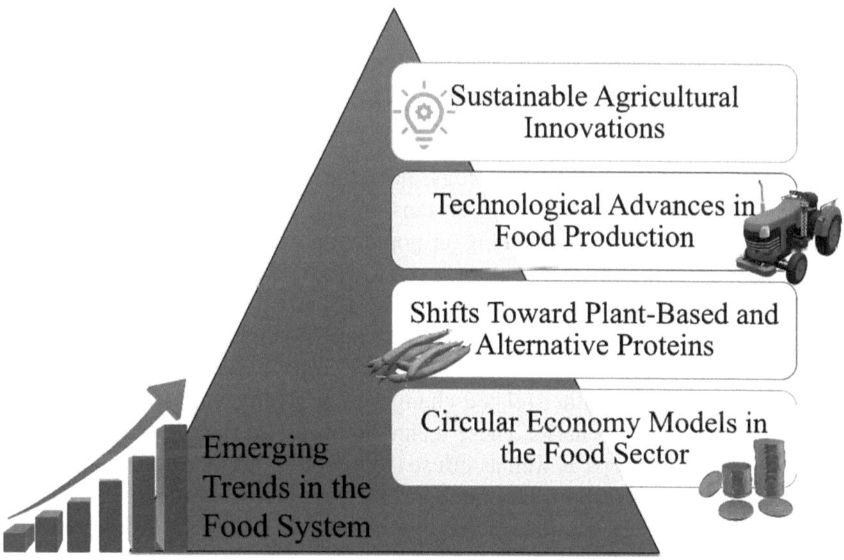

Fig. 5.2 Emerging trends in food system

most promising innovations in sustainable agriculture is precision farming, which makes use of advanced technologies including drones, satellites and soil sensors to manage crops management in more efficient way. This also enables water, fertilizers, and pesticides to be applied in targeted manners to save waste and minimize negative impacts on the environment. Precision agriculture conducts targeted interventions to specific areas of the field, thus significantly reducing input costs and conserving much-needed natural resources, very basic in a world with increasingly stringent environmental constraints (Kanannavar et al., 2020). The other trend being adopted is the regenerative agriculture approach mentioned in Table 5.1. Here, along with this, the main focus is to bring back soil health, increase biodiversity, and sequester carbon, which is something different from conventional farming where soils are depleted and biodiversity is almost negligible, but it also happens to be farmed for its proven ability to produce good quality nutritious food with fewer inputs and with fewer externalities. These methods include cover cropping, minimal tillage, and rotational grazing that also allow for greater nutrient- and water-retentive capabilities in soils while diminishing soil erosion. Further, rotational grazing locks carbon, which further aids in mitigating climate change. Companies and governments throughout the world are increasingly embracing regenerative agriculture because it's the future of truly sustainable agriculture (Specht et al., 2015). Agroecology supports and promotes productive, sustainable agricultural systems. It is an approach that combines ecological principles for farming practice, emphasizing the relationships between organisms and their environment. Agroecology values biodiversity; it encourages natural pest control and reduced use of chemical inputs, improves traditional knowledge, inspires farmers to collaborate with their local ecosystems, that are resilient against climate change, and raises yields while increasing food security (Powlson et al., 2011).

5.3.2 Technological Innovations in Food Production

Technologies ranging from biotechnology, and robotics, to artificial intelligence have all advanced at warp speed, and this is transforming food production; a move that will lead to more food for a much bigger population while negative impacts on the environment are minimized. The most promising advance comes in the domain of genetic editing, and, among others, with tools such as CRISPR—a new dimension of plant genome modification that is entirely focused on highly precise interventions into the crop's genome. The potential to engineer crops that will be less susceptible to disease, pests, and weather-related changes may greatly improve food security in the climate change scenario. These technologies are to raise yields and reduce chemical inputs into crops as well as infuse them with better nutrition (Walait et al., 2024). Automation and robotics further boost food production productivity. The said automation and robotics can plant seeds, weed out grass and other overgrowth, and harvest crops with much greater accuracy and greater efficiency while considerably reducing labor costs along with increased productivity. As an example, AI-based GPS technology for independent tractors and harvesters can run 24/7 to maximize crop

Table 5.1 Understanding how various communities approach sustainable food practices

Study areas	Primary problem	Objective	A sustainable approach to resolving issues	References
Nutrition	Nutrition debates have overlooked the extremely fast changes occurring in the food system. Undernutrition and overnutrition, which result from the double burden of malnutrition, are two of the most significant nutrition issues facing food systems	Provide everyone with healthy, secure, and easily available food	To guarantee adequate nutritional intake, promote variety in the consumption of fruits, vegetables, whole grains, and legumes Expand local mechanisms for food production and distribution to improve access to wholesome, fresh foods	Béné et al. (2018) Herforth and Ahmed (2015)
Agriculture	The long-term viability of food production is threatened by the degradation and depletion of the natural assets that support the food system. Long-term intensive farming, monocropping, and excessive use of chemical fertilizers have all contributed to soil deterioration, which lowers crop yields and makes the soil more susceptible to pests and diseases	Enhance soil vitality, improve crop resilience, and optimize water utilization	Regenerative agriculture, precision agriculture, crop rotation, and conservation tillage The sustainable expansion of agriculture is defined as a "valuable guiding principle" for harmonizing sustainability with production	Dubé et al. (2014)

(continued)

Table 5.1 (continued)

Study areas	Primary problem	Objective	A sustainable approach to resolving issues	References
Environmental and social issues	Global environmental change, with social, political, and economic transformations, has imposed unparalleled pressures on food security and agricultural systems	Support social equity in food distribution	It is necessary to do a more thorough and complete investigation of how the way food production, processing, distribution, and consumption are currently organized to support food security. To do this, the idea of a "food system" must be expanded beyond its operations to take into account the system's social, economic, and ecological effects and suggestions	Ericksen (2007)
Engineering	Efficiency in transportation, storage, and production	Build technologies for storing and producing food in a sustainable manner	Innovations in cold storage, IoT in farming, and precision agricultural technologies	Awulachew (2023)
Agroecology	The problem of contemporary, extractive agriculture, mostly characterized by monoculture, has resulted in significant environmental damage, including loss of biodiversity, loss of soil nutrients, and pollution, alongside the exclusion of the small-scale farmer	Mitigation of climate change and encouragement of adaptation practices Climate-resilient agriculture	Emphasize the role of humans as integral components of agroecosystems and recognize natural capital as a vital ingredient for maintaining agricultural output, as well as how well agriculture fits in with other ecological functions and processes	Eakin et al. (2016), Gliessman (2021)

management while reducing fuel usage. This will be one of the greatest responses to this growing issue concerning labor scarcity in most parts of the world today (Kaur et al., 2023). Indeed, AI and big data are revolutionizing food production management. AI-based systems can scan the humongous data from sensors, drones, and weather to predict crop yields, monitor soil health, and detect diseases early, and all this would translate into efficient operations with fewer wastes in matters such as when to plant, water, or harvest. AI also helps improve supply chains to transport food and share it in the most cost- and environmentally-friendly manner possible (Kothari, 2024).

5.3.3 *Growth in Plant-Based and Alternative Proteins*

Growing awareness of meat consumption's environmental and health impacts has sparked tremendous demand for plant-based diets and alternative proteins. Meat and dairy production consume water, land, and feed; they emit greenhouse gases and contribute to the gobbling up of the world's forests. Consumers are choosing more sustainable plant-based alternatives with reduced environmental footprints. Those plant-based protein products derived from soy and peas in comparison to other legumes show much promise, given the growth in the flexitarian, vegetarian, or vegan diet forms taken up by different consumers. They are made to mimic in taste, texture, and appearance, as more environmentally conscious (Tachie et al., 2023). While plant-based and lab-grown meats are perhaps the most promising sources of nutrition since the early days, insect protein could arguably be considered another form of promising source of nutrition (Rubio et al., 2020). Algae and microbial proteins have also emerged as promising alternative sources of traditional animal and plant-based proteins as represented in Table 5.2. Algae such as spirulina and chlorella have high nutritional content and could be farmed on water bodies that may not be suitable for normal agriculture. The same is to be said about microbial proteins derived from fungi: here, sustainable protein with minimal environmental footprint is available. From the famous protein bars to meat replacers, algae, and microbial proteins are entering wide range of foods, opening new vistas in sustainable diets (Ahmad & Ashraf, 2023).

5.3.4 *Circular Economy Models in the Food Sector*

The latest concepts picked ground in the food sector are described as reducing waste, conserving resources, and making a more sustainable food system. While resources, using them, and discarding them define the traditional model of a linear economy, circular economies aim at closing the loop to design out waste and keep materials

Table 5.2 Comparison of the nutritional quality of alternative proteins

Protein type	Source	Protein content (per 100 g)	Essential amino-acid content (per 100 g)	Fat content %	References
Plant protein	Soy	14.3–38	32.4	7.7–47	Messina (1999)
	Almonds	25.8	40.08	39–57	House et al. (2019)
	Walnuts	13–25	30.08	69–80	Kafkas et al. (2020)
	Oats	14.4	36.04	78–81.5	Biel et al. (2009), Sterna et al. (2016)
	Brown rice	12	38.93	2.63	Longvah et al. (2021)
	Moringa oleifera leaves	24	40.29	12.5–21.6	Yaméogo et al. (2011)
Algal protein	*Chlorella vulgaris*	53.5	39.45	41.9–66.3	Maurício et al. (2023), Wang et al. (2020)
Insect protein	House cricket	54	36.95	34.51	Köhler et al. (2019)
Myco-protein	*Fusarium venenatum*	52	40.80	3.5	Zhang et al. (2019)

in use for as long as possible (Rabbi & Amin, 2024). It appears that food reduction waste is at the epicenter of many aspects of the circular economy. Improvements such as enhanced storage and preservation methods as well as redistribution networks could ensure that more food reaches the consumer and less becomes waste. Another means to eliminate waste landing in landfills would be to convert wasted food into new products such as fruit peels and vegetable scraps as natural sweeteners and compost, respectively (Patyal et al., 2022). Resource recycling in agriculture, therefore, becomes the other core component of the circular economy. Currently, farmers are embracing high-intensity farming systems for example water and nutrient recycling to reduce waste and increase efficiency in farming (Awino & Apitz, 2023).

5.4 Policy and Governance for Sustainable Food Systems

The governments recognize that there is a need for practices that have placed environmental conservation, economic viability, and social equity as balanced factors in food security. Government subsidies have been part of agricultural policy. Such subsidies for input materials such as fertilizers, pesticides, and water have helped farmers to produce increased productivity along with ensuring food security (Galli et al., 2020). International and regional food markets are highly impacted by trade policies. Traditionally, protectionist policies pursued in the name of supporting the domestic agricultural sector often resulted in market inefficiencies, higher food prices, and restricted access to a diversified source of food (Willett et al., 2019). Globalization and liberalization policies were advanced to encourage more efficient markets and greater food movements across borders. It presents a challenge too: market price volatility threatens small-scale farmers, and exploitation in global supply chains persists (Caron et al., 2018). The kind of loss suffered may see every likelihood that sustainable trade policies may now be put in focus. Emphasis has been placed on balancing domestic food security commitment with international trade commitments ensuring fair market access to all concerned, with due recognition of environmental and social considerations (Fanzo, 2021). As environmental concern keeps growing, so has the attention of governments on sustainable agricultural policies. The policies may also enhance agroforestry, precision agriculture, and organic farming in an attempt to reduce the ecological footprint of agriculture. Environmental issues are increasingly becoming a preference in most policy frameworks, and this now contributes to an upward curve. Several agricultural development programs are out there to minimize chemical inputs, bolster soil health, and lessen effects linked to climate risks (Willett et al., 2019). At times, policies have catalyzed inequalities, as they instead tipped the balance in favor of large commercial farms versus smallholders. Distributions of resources and opportunities have remained some of the persisting problems, characterized by gender imbalances. These are manifested in more modern policies of social equity and inclusion. Some of these include land reform programs, targeted support for specific marginalized communities, and the empowerment of women, among others. Governments increasingly realize the importance of having inclusive policies to redress the differentiated needs of all in the food system (Webb et al., 2020).

5.4.1 *Environmental Conservation*

Several governments have anchored their policies on sustainable food systems with agroecological practices. These include the reduction of the ecological footprint of agriculture, and enhancement of soil health and biodiversity, including the practices on crop rotation, polyculture, and natural pest control. The government is even more invested in climate-smart agricultural practices, where farming systems become even more resilient to climate change. Under this falls the development of drought-resistant crops, bettering the management of water, and promotion of sustainable land-use practices (Friel et al., 2020). Conservation of Ecosystems conservation efforts, such as reforestation and biodiversity protection hotspots, are therefore included in the greater efforts in the creation of sustainable food systems. Those have realized that agriculture is connected with the larger aspect healthy environment (Rockström et al., 2020).

5.4.2 Economic Resilience

The Sustainability of the economy will come from support given by the government to small-scale farmers. Programs that provide credit training, and opening up access to markets empower local producers while making them better equipped to withstand the economic shocks and promote diversified food production. Market Access and Fair Trade in respect of market access policies that make things easy and efforts at fair trade, contribute to the economy (Ng, 2018). The economy with a good price paid for the produce means food can build local economies and encourage responsible sourcing in global supply chains. Investing in rural infrastructures would establish the economic sustainability of food systems. These include transport networks and storage facilities. Improved infrastructure ensures efficient movements of goods whereby post-harvest losses are reduced and the overall effect on the economy has enhanced (Edwards et al., 2024).

5.5 Future Pathways for a Sustainable Food System

5.5.1 Precision Agriculture

More efficient use of resources, increased productivity and lower environmental impacts are expected to result from the application of precision farming technologies, such as drones, sensors, and the analytics of big data. The governments will be able to motivate and encourage smallholder farms for the use of the technologies through training and subsidizing the products (Nhamo et al., 2024). The proper application of biotechnology offers the possibility of developing crop varieties that can be tolerant to changing climatic conditions as well as enhance the nutritional quality of the produce. The governments shall encourage developments in agricultural biotechnology by funding research projects but at the same time observing severe levels of safety testing and openness to the public about the threats posed by such technologies (Searchinger et al., 2022).

5.5.2 Climate-Smart Agriculture

Improvement in the areas of research and development on drought-resilient crop varieties would help counter the effects of water scarcity on agricultural production. Governments may encourage the adoption of climate-smart agriculture through provision of resilient seeds and sustainable farming practices to farmers (Lipper et al., 2014). One of the opportunities for using agricultural landscapes is agroforestry. Governments can make incentives that encourage growing a number of trees on the farm to increase biodiversity, improve soil health, and sequester carbon (Antonelli, 2023).

5.5.3 Sustainable Supply Chains

This would promote localized and regional food systems and reduce dependency on global supply chains, thereby making the food systems more resilient to disruption. Governments can undertake initiatives that strengthen local food networks, for example, farmers' markets, community-supported agriculture, and farm-to-table programs (Accorsi et al., 2019). Supply chains enhanced by block chain technology increase transparency and traceability. The innovation may contribute to the reduction of food safety, fair trade, and sustainability concerns. With engagement with the private sector, governments can step out to implement block chain solutions in food supply chains (Bhat & Jõudu, 2019).

5.5.4 Policy Integration and Coherence

Needless to say, there is a requirement for an integrated approach to such issues that encompasses all policies in all sectors such as agriculture, environment, health, and trade. Policy coherence in relation to these interrelated issues can be promoted by establishing interdepartmental task forces (Edwards et al., 2024). It is possible for governments to design systems of incentives that promote sustainability. This may include grants for eco-friendly agricultural practices, tax breaks to firms for adopting circular economy models, and financing social equity projects in food systems (Tosun & Lang, 2017).

5.5.5 Education and Training

Farmers require integrated training programs that focus on best practices in farming, mobilizing technology, and dealing with climate change impact. The government can join hands with educational institutions and extension services in providing effective and customizable trainings (Li et al., 2015). Besides, the governments can place advertisements to sensitize the public on the environmental and social effects of eating certain foods. Culture changes that promote consumption of healthy and sustainable food should increase the supply of responsibly produced food in the market, which will, in turn, influence the competition in markets (Paxman & Nield, 2022).

5.5.6 International Collaboration

There is a need for international collaboration in knowledge sharing and capacity enhancement. It would be in a position to conduct and implement collaborative research, disseminate the best practices and tackle global problems like climate change more effectively. Create a priority trade policy partnership among the states. It includes the promotion of sustainable development, support for social trade agreements, and protection of the society and environment in the course of trade negotiations (Melkonyan et al., 2019).

5.6 Conclusion

We need to change our perspective to address the challenges of the Anthropocene. We should acknowledge it as a set of socioecological systems that require change rather than just as a collection of concerning numbers. The agri-food system provides a distinctive viewpoint for critically discussing the idea of the Anthropocene. By focusing on food and agriculture, we may observe the inequalities and variety that characterize each proposed chronological start of this human-dominated age. Food shortages are exacerbated by high-input agricultural methods that degrade natural resources, disturb the ecological system, and by weather patterns that cast uncertainty on what is produced. Poor infrastructure and economic inequalities also limit people's access to nutrients that might otherwise prevent malnutrition and health issues. This increased interest in plant-based and alternative proteins represents a shift in the paradigm of consumer preferences, based on growing awareness of the environmental implications of meat production. It supports more sustainable diets while creating new opportunities for economic growth and innovation in food production. The principles of circular economy in the food sector highlight the need to reduce waste and increase resource efficiency, thereby providing a holistic approach to addressing systemic inefficiencies. Gradual progress can be shown by a farmer introducing a climate-smart crop, a county-level decision maker investing in improved extension services, or a food systems researcher creating a more spatially explicit, integrated model that connects food systems, ecosystem services, and urban land use to assist government planners. These "small wins," "success stories of closing loops," and "bright seeds" show how regional initiatives can result in major advancements in the food chain. There needs to be coordination among the stakeholders including farmers, policymakers, researchers, and consumers. The food system must function as we move forward with this level of stakeholder coordination so that it not only survives but flourishes in the face of every difficulty of the Anthropocene.

References

Accorsi, R., Ferrari, E., & Manzini, R. (2019). Modeling inclusive food supply chains toward sustainable ecosystem planning. In *Elsevier eBooks* (pp. 1–21). https://doi.org/10.1016/b978-0-12-813411-5.00001-6

Ahmad, A., & Ashraf, S. S. (2023). Sustainable food and feed sources from microalgae: Food security and the circular bioeconomy. *Algal Research, 74*, Article 103185. https://doi.org/10.1016/j.algal.2023.103185

Alaimo, K., Chilton, M., & Jones, S. J. (2020). Food insecurity, hunger, and malnutrition. In *Elsevier eBooks* (pp. 311–326). https://doi.org/10.1016/b978-0-12-818460-8.00017-4

Antonelli, A. (2023). Indigenous knowledge is key to sustainable food systems. *Nature, 613*(7943), 239–242. https://doi.org/10.1038/d41586-023-00021-4

Awino, F. B., & Apitz, S. E. (2023). Solid waste management in the context of the waste hierarchy and circular economy frameworks: An international critical review. *Integrated Environmental Assessment and Management, 20*(1), 9–35. https://doi.org/10.1002/ieam.4774

Awulachew, M. T. (2023). Employing engineering principles and business model innovation in food systems to achieve sustainable development goals. *Agronomy & Agricultural Science, 6*(1), 1–8. https://doi.org/10.24966/aas-8292/100046

Bhat, R., & Jõudu, I. (2019). Emerging issues and challenges in agri-food supply chain. In *Elsevier eBooks* (pp. 23–37). https://doi.org/10.1016/b978-0-12-813411-5.00002-8

Biel, W., Bobko, K., & Maciorowski, R. (2009). Chemical composition and nutritive value of husked and naked oats grain. *Journal of Cereal Science, 49*(3), 413–418. https://doi.org/10.1016/j.jcs.2009.01.009

Von Braun, J. (2010). Food insecurity, hunger and malnutrition: Necessary policy and technology changes. *New Biotechnology, 27*(5), 449–452. https://doi.org/10.1016/j.nbt.2010.08.006

Béné, C., Oosterveer, P., Lamotte, L., Brouwer, I. D., De Haan, S., Prager, S. D., Talsma, E. F., & Khoury, C. K. (2018). When food systems meet sustainability—Current narratives and implications for actions. *World Development, 113*, 116–130. https://doi.org/10.1016/j.worlddev.2018.08.011

Caron, P., De Loma-Osorio, G. F. Y., Nabarro, D., Hainzelin, E., Guillou, M., Andersen, I., Arnold, T., Astralaga, M., Beukeboom, M., Bickersteth, S., Bwalya, M., Caballero, P., Campbell, B. M., Divine, N., Fan, S., Frick, M., Friis, A., Gallagher, M., Halkin, J., et al. (2018). Food systems for sustainable development: Proposals for a profound four-part transformation. *Agronomy for Sustainable Development, 38*(4). https://doi.org/10.1007/s13593-018-0519-1

Dubé, L., Webb, P., Arora, N. K., & Pingali, P. (2014). Agriculture, health, and wealth convergence: Bridging traditional food systems and modern agribusiness solutions. *Annals of the New York Academy of Sciences, 1331*(1), 1–14. https://doi.org/10.1111/nyas.12602

Eakin, H., Connors, J. P., Wharton, C., Bertmann, F., Xiong, A., & Stoltzfus, J. (2016). Identifying attributes of food system sustainability: Emerging themes and consensus. *Agriculture and Human Values, 34*(3), 757–773. https://doi.org/10.1007/s10460-016-9754-8

Edwards, F., Sonnino, R., & Cifuentes, M. L. (2024). Connecting the dots: Integrating food policies towards food system transformation. *Environmental Science & Policy, 156*, Article 103735. https://doi.org/10.1016/j.envsci.2024.103735

Ericksen, P. J. (2007). Conceptualizing food systems for global environmental change research. *Global Environmental Change, 18*(1), 234–245.

Fanzo, J. (2021). Achieving equitable diets for all: The long and winding road. *One Earth, 4*(4), 470–473. https://doi.org/10.1016/j.oneear.2021.03.007
Finney, S. C., & Edwards, L. E. (2016). The "Anthropocene" epoch: Scientific decision or political statement? *GSA Today*, 26(3), 4–10. https://doi.org/10.1130/GSATG270A.1
Fresco, L. O. (2009). Challenges for food system adaptation today and tomorrow. *Environmental Science & Policy, 12*(4), 378–385. https://doi.org/10.1016/j.envsci.2008.11.001
Friel, S., Schram, A., & Townsend, B. (2020). The nexus between international trade, food systems, malnutrition and climate change. *Nature Food, 1*(1), 51–58. https://doi.org/10.1038/s43016-019-0014-0
Galli, F., Prosperi, P., Favilli, E., D'Amico, S., Bartolini, F., & Brunori, G. (2020). How can policy processes remove barriers to sustainable food systems in Europe? Contributing to a policy framework for agri-food transitions. *Food Policy, 96*, Article 101871. https://doi.org/10.1016/j.foodpol.2020.101871
Gibbard, P., Walker, M., Bauer, A., Edgeworth, M., Edwards, L., Ellis, E., Finney, S., Gill, J. L., Maslin, M., Merritts, D., & Ruddiman, W. (2022). The Anthropocene as an Event, not an Epoch. *Journal of Quaternary Science*, 37(3), 395–399. https://doi.org/10.1002/jqs.3416
Gliessman, S. R. (2021). *Package price agroecology: The ecology of sustainable food systems*. CRC Press.
Gomez-Zavaglia, A., Mejuto, J., & Simal-Gandara, J. (2020). Mitigation of emerging implications of climate change on food production systems. *Food Research International, 134*, Article 109256. https://doi.org/10.1016/j.foodres.2020.109256
Grumbine, R. E., Xu, J., & Ma, L. (2021). An overview of the problems and prospects for circular agriculture in sustainable food systems in the Anthropocene. *Circular Agricultural Systems, 1*(1), 1–11. https://doi.org/10.48130/cas-2021-0003
Gupta, G. S. (2019). Land degradation and challenges of food security. *Review of European Studies, 11*(1), 63. https://doi.org/10.5539/res.v11n1p63
Herforth, A., & Ahmed, S. (2015). The food environment, its effects on dietary consumption, and potential for measurement within agriculture-nutrition interventions. *Food Security, 7*(3), 505–520. https://doi.org/10.1007/s12571-015-0455-8
Hird, M. J., & Yusoff, K. (2016). Subtending relations: Bacteria, geology, and the possible. In *Genealogies of speculation: Materialism and subjectivity since structuralism* (pp. 319–342).
Hossain, A., Krupnik, T. J., Timsina, J., Mahboob, M. G., Chaki, A. K., Farooq, M., Bhatt, R., Fahad, S., & Hasanuzzaman, M. (2020b). Agricultural land degradation: Processes and problems undermining future food security. In *Springer eBooks* (pp. 17–61). https://doi.org/10.1007/978-3-030-49732-3_2
House, J. D., Hill, K., Neufeld, J., Franczyk, A., & Nosworthy, M. G. (2019). Determination of the protein quality of almonds (Prunus dulcis L.) as assessed by in vitro and in vivo methodologies. *Food Science & Nutrition*, 7(9), 2932–2938. https://doi.org/10.1002/fsn3.1146
Kafkas, E., Attar, S. H., Gundesli, M. A., Ozcan, A., & Ergun, M. (2020). Phenolic and fatty acid profile, and protein content of different walnut cultivars and genotypes (Juglans regia L.) grown in the USA. *International Journal of Fruit Science, 20*(sup3), S1711–S1720. https://doi.org/10.1080/15538362.2020.1830014
Kanannavar, P. S., Premanand, B. D., Subhas, B., Anuraja, B., & Bhogi, P. B. (2020). Laser land levelling—An engineering approach for scientific irrigation water management in irrigation command areas of Karnataka, India. *International Journal of Current Microbiology and Applied Sciences, 9*(5), 2393–2398. https://doi.org/10.20546/ijcmas.2020.905.272
Kaur, K., Priyanka, N., Kaur, G., Singh, B., Sehgal, S., & Trehan, S. (2023). Artificial Intelligence (AI) as a transitional tool for sustainable food systems. In *World sustainability series* (pp. 305–328). https://doi.org/10.1007/978-3-031-46046-3_15
Kothari, H. (2024). Revolutionary trends in food and technology. In *Futuristic trends in agriculture engineering & food sciences* (Vol. 3, Book 1, pp. 88–95). https://doi.org/10.58532/v3bcagp1ch6

Köhler, R., Kariuki, L., Lambert, C., & Biesalski, H. (2019). Protein, amino acid and mineral composition of some edible insects from Thailand. *Journal of Asia-Pacific Entomology*, 22(1), 372–378. https://doi.org/10.1016/j.aspen.2019.02.002

Lewis, S. L., & Maslin, M. A. (2015). Defining the Anthropocene. *Nature*, 519(7542), 171–180. https://doi.org/10.1038/nature14258

Li, H., Wu, T., Wang, X., & Qi, Y. (2015). The greenhouse gas footprint of China's food system: An analysis of recent trends and future scenarios. *Journal of Industrial Ecology, 20*(4), 803–817. https://doi.org/10.1111/jiec.12323

Lipper, L., Thornton, P., Campbell, B. M., Baedeker, T., Braimoh, A., Bwalya, M., Caron, P., Cattaneo, A., Garrity, D. P., Henry, K., Hottle, R., Jackson, L., Jarvis, A., Kossam, F., Mann, W., McCarthy, N., Meybeck, A., Neufeldt, H., Remington, T., et al. (2014). Climate-smart agriculture for food security. *Nature Climate Change, 4*(12), 1068–1072. https://doi.org/10.1038/nclimate2437

Longvah, T., Mangthya, K., Subhash, K., Sen, S., & Rathi, S. (2021). Comprehensive nutritional evaluation of popular rice varieties of Assam, Northeast India. *Journal of Food Composition and Analysis, 101*, Article 103952. https://doi.org/10.1016/j.jfca.2021.103952

Lougheed, S. C., & Hird, M. J. (2017). Food security and secure food in the Anthropocene. *Crime Law and Social Change, 68*(5), 499–514. https://doi.org/10.1007/s10611-017-9699-x

Maurício, T., Couto, D., Lopes, D., Conde, T., Pais, R., Batista, J., Melo, T., Pinho, M., Moreira, A. S. P., Trovão, M., Barros, A., Cardoso, H., Silva, J., Domingues, P., & Domingues, M. R. (2023). Differences and Similarities in Lipid Composition, Nutritional Value, and Bioactive Potential of Four Edible Chlorella vulgaris Strains. *Foods*, 12(8), 1625. https://doi.org/10.3390/foods12081625

Mazac, R., & Tuomisto, H. L. (2020). The post-Anthropocene diet: Navigating future diets for sustainable food systems. *Sustainability, 12*(6), 2355. https://doi.org/10.3390/su12062355

Melkonyan, A., Gruchmann, T., Huerta, A., & Krumme, K. (2019). Scenario planning for sustainable food supply chains. In *Springer eBooks* (pp. 211–233). https://doi.org/10.1007/978-3-319-98467-4_10

Messina, M. J. (1999). Legumes and soybeans: Overview of their nutritional profiles and health effects. *American Journal of Clinical Nutrition, 70*(3), 439S-450S. https://doi.org/10.1093/ajcn/70.3.439s

Mrabet, R. (2023). Sustainable agriculture for food and nutritional security. In *Elsevier eBooks* (pp. 25–90). https://doi.org/10.1016/b978-0-323-90500-8.00013-0

Ng, A. W. (2018). From sustainability accounting to a green financing system: Institutional legitimacy and market heterogeneity in a global financial centrer. *Journal of Cleaner Production, 195*, 585–592. https://doi.org/10.1016/j.jclepro.2018.05.250

Nhamo, L., Mpandeli, S., Liphadzi, S., Hlophe-Ginindza, S., & Mabhaudhi, T. (2024). Transitional pathways towards sustainable food systems. In *CRC Press eBooks* (pp. 60–77). https://doi.org/10.1201/9781003327615-4

Patyal, V. S., Sarma, P., Modgil, S., Nag, T., & Dennehy, D. (2022). Mapping the links between Industry 4.0, circular economy and sustainability: A systematic literature review. *Journal of Enterprise Information Management, 35*(1), 1–35. https://doi.org/10.1108/jeim-05-2021-0197

Paxman, J. R., & Nield, L. (2022). Editorial: food and nutrition: Pathways to a sustainable future. *Proceedings of the Nutrition Society*, 1–4. https://doi.org/10.1017/s0029665122002865

Powlson, D. S., Whitmore, A. P., & Goulding, K. W. T. (2011). Soil carbon sequestration to mitigate climate change: A critical re-examination to identify the true and the false. *European Journal of Soil Science, 62*(1), 42–55. https://doi.org/10.1111/j.1365-2389.2010.01342.x

Rabbi, M. F., & Amin, M. B. (2024). Circular economy and sustainable practices in the food industry: A comprehensive bibliometric analysis. *Cleaner and Responsible Consumption, 14*, Article 100206. https://doi.org/10.1016/j.clrc.2024.100206

Reisman, E., & Fairbairn, M. (2020). Agri-food systems and the Anthropocene. *Annals of the American Association of Geographers, 111*(3), 687–697. https://doi.org/10.1080/24694452.2020.1828025

Rockström, J., Williams, J., Daily, G., Noble, A., Matthews, N., Gordon, L., Wetterstrand, H., DeClerck, F., Shah, M., Steduto, P., De Fraiture, C., Hatibu, N., Unver, O., Bird, J., Sibanda, L., & Smith, J. (2016). Sustainable intensification of agriculture for human prosperity and global sustainability. *Ambio, 46*(1), 4–17. https://doi.org/10.1007/s13280-016-0793-6

Rockström, J., Edenhofer, O., Gaertner, J., & DeClerck, F. (2020). Planet-proofing the global food system. *Nature Food*, 1(1), 3–5. https://doi.org/10.1038/s43016-019-0010-4

Rubio, N. R., Xiang, N., & Kaplan, D. L. (2020). Plant-based and cell-based approaches to meat production. *Nature Communications, 11*(1). https://doi.org/10.1038/s41467-020-20061-y

Scherer, L., Svenning, J., Huang, J., Seymour, C. L., Sandel, B., Mueller, N., Kummu, M., Bekunda, M., Bruelheide, H., Hochman, Z., Siebert, S., Rueda, O., & Van Bodegom, P. M. (2020). Global priorities of environmental issues to combat food insecurity and biodiversity loss. *The Science of the Total Environment, 730*, Article 139096. https://doi.org/10.1016/j.scitotenv.2020.139096

Searchinger, T., James, O., Dumas, P., Kastner, T., & Wirsenius, S. (2022). EU climate plan sacrifices carbon storage and biodiversity for bioenergy. *Nature, 612*(7938), 27–30. https://doi.org/10.1038/d41586-022-04133-1

Specht, K., Siebert, R., Thomaier, S., Freisinger, U., Sawicka, M., Dierich, A., Henckel, D., & Busse, M. (2015). Zero-acreage farming in the city of Berlin: An aggregated stakeholder perspective on potential benefits and challenges. *Sustainability, 7*(4), 4511–4523. https://doi.org/10.3390/su7044511

Sterna, V., Zute, S., & Brunava, L. (2016). Oat grain composition and its nutrition benefice. *Agriculture and Agricultural Science Procedia, 8*, 252–256. https://doi.org/10.1016/j.aaspro.2016.02.100

Tachie, C., Nwachukwu, I. D., & Aryee, A. N. A. (2023). Trends and innovations in the formulation of plant-based foods. *Food Production Processing and Nutrition, 5*(1). https://doi.org/10.1186/s43014-023-00129-0

Thomas, C. D. (2022). Maintaining global biodiversity by developing a sustainable Anthropocene food production system. *The Anthropocene Review, 9*(3), 379–391. https://doi.org/10.1177/20530196221129747

Tosun, J., & Lang, A. (2017). Policy integration: Mapping the different concepts. *Policy Studies, 38*(6), 553–570. https://doi.org/10.1080/01442872.2017.1339239

Walait, M., Ajaz, M., Rasool, W., Irfan, M., Fatima, M., & Tariq, F. (2024). Biotechnological innovations in sustainable food production. *DIET FACTOR (Journal of Nutritional & Food Sciences)*, 02–09. https://doi.org/10.54393/df.v5i01.107

Wang, Y., Tibbetts, S. M., Berrue, F., McGinn, P. J., MacQuarrie, S. P., Puttaswamy, A., Patelakis, S., Schmidt, D., Melanson, R., & MacKenzie, S. E. (2020). A Rat Study to Evaluate the Protein Quality of Three Green Microalgal Species and the Impact of Mechanical Cell Wall Disruption. *Foods*, 9(11), 1531. https://doi.org/10.3390/foods9111531

Webb, P., Benton, T. G., Beddington, J., Flynn, D., Kelly, N. M., & Thomas, S. M. (2020). The urgency of food system transformation is now irrefutable. *Nature Food, 1*(10), 584–585. https://doi.org/10.1038/s43016-020-00161-0

Willett, W., Rockström, J., Loken, B., Springmann, M., Lang, T., Vermeulen, S., Garnett, T., Tilman, D., DeClerck, F., Wood, A., Jonell, M., Clark, M., Gordon, L. J., Fanzo, J., Hawkes, C., Zurayk, R., Rivera, J. A., De Vries, W., Sibanda, L. M., et al. (2019). Food in the Anthropocene: The EAT–Lancet Commission on healthy diets from sustainable food systems. *The Lancet, 393*(10170), 447–492. https://doi.org/10.1016/s0140-6736(18)31788-4

Yaméogo, C. W., Bengaly, M. D., Savadogo, A., Nikiema, P. A., & Traore, S. A. (2011). Determination of chemical composition and nutritional values of Moringa oleifera leaves. *Pakistan Journal of Nutrition, 10*(3), 264–268.

Zalasiewicz, J., Williams, M., Haywood, A., & Ellis, M. (2011). The Anthropocene: a new epoch of geological time? *Philosophical Transactions of the Royal Society A: Mathematical, Physical and Engineering Sciences*, 369(1938), 835–841. https://doi.org/10.1098/rsta.2010.0339

Zhang, J., Li, X., Ying, Y., & Yao, X. (2019). Effects of the Castanea mollissima Blume Shell Cultivation Substrate on the Yield and Nutritional Composition of Pleurotus geesteranus. *Food and Nutrition Sciences,10*(09), 1112–1121. https://doi.org/10.4236/fns.2019.109080

Chapter 6
Challenges and Innovations in Food Safety: Addressing Microbiome, Pathogens, and Global Risks

Adeyemi T. Odudimu, Ruth C. Ebubechukwu, Muyideen O. Bamidele, Ezekiel O.Faluyi, Bamikale M. Bola, and Chidiebere I. Nwaogbo

Abstract Throughout human history, food has been fundamental to survival. However, unsafe food poses significant health risks, causing approximately 420,000 deaths annually, according to the World Health Organization. Those most affected are in developing and underdeveloped nations. Food safety issues often arise from food microbiome and foodborne pathogens. Microbiome, which include communities of fungi, bacteria, and viruses, interact with different food environments—raw, processed, and packaged—altering their sensory properties. This microbiome can either prevent harmful pathogens or contribute to foodborne diseases. This chapter examines food safety challenges such as managing foodborne pathogens, antimicrobial resistance, detecting microbiome, and the impacts of climate change. Other topics include food preservation, harmful microorganisms, increased contamination risks due to globalization, government regulations, food packaging, technological advancements, and the costs of control. Various scientific innovations by food scientists aim to address these challenges. The chapter concludes with a discussion of these innovations and their potential to improve food safety.

A. T. Odudimu · R. C. Ebubechukwu
Department of Chemistry, Lehigh University, Bethlehem Pennsylvania, USA
e-mail: ado219@lehigh.edu

R. C. Ebubechukwu
e-mail: rce221@lehigh.edu

M. O. Bamidele (✉) · E. O.Faluyi · B. M. Bola
Department of Food Research, School of Chemistry, Autonomous University of Coahuila, Saltillo, Mexico
e-mail: mbamidele@uadec.edu.mx

E. O.Faluyi
e-mail: e_faluyi@uadec.edu.mx

B. M. Bola
e-mail: bamikalemicheal@uadec.edu.mx

C. I. Nwaogbo
Department of Physics, Lehigh University, Bethlehem Pennsylvania, USA
e-mail: cin221@lehigh.edu

J. Parray and A. K. Haghi (eds.), *Food and Water Security*,
https://doi.org/10.1007/978-3-032-04870-7_6

Keywords Food safety · Microbiome · Foodborne pathogens · Nanotechnology innovations · Prebiotics and probiotics

6.1 Introduction

Even though food is a major basic need of not just humans but also all living creatures, maintaining and ensuring the safety of food substances is as important as the food itself. As the world population grows, there is an increasing need for innovations to enact food safety.

According to the (World Health Organization, 2022) factsheet, 1 out of every 10 persons are hospitalized because of food contamination, 600 million people globally, and 420,000 deaths are recorded among the cases annually. In addition, an alarming record of a financial burden of about US$ 100 billion lost (World Health Organization, 2022).

6.2 Classification of Food Microbiome

Food microbiome are ubiquitous, so we group them under two classes (Fig. 6.1).

6.2.1 Classification Based on Food Environment

i. **Human association**

The human body can harbor a community of microorganisms and function as a healthy carrier for several foodborne pathogens. For instance, food poisoning caused by *Escherichia coli* can occur due to food in contact with human feces. This and many other food contaminations can arise due to association with human body fluid (Aggarwal et al., 2023; Dekaboruah et al., 2020).

ii. **Soil**

Soil microbiome can both be beneficial and detrimental to food safety. In the former, they promote plant growth due to symbiotic relationships, which aid nitrogen fixation and nutrient acquisition in the soil, for example, a study showed that the presence of bacteria like *Achrobacter denitrificans, Pseudomonas entremorientalis, and Rhizobium pusenser* in the soil increases the growth of mungbean (Zahra et al., 2023). In the latter, the soil can harbor disease-causing agents that can cause plant diseases or are a source of opportunistic infections in humans and animals through crop contamination (Blum et al., 2019), notable examples are root-knot nematodes, which causes root-knot galls in plants (Rutter et al., 2022), *Rhizopus species* grows in soil and

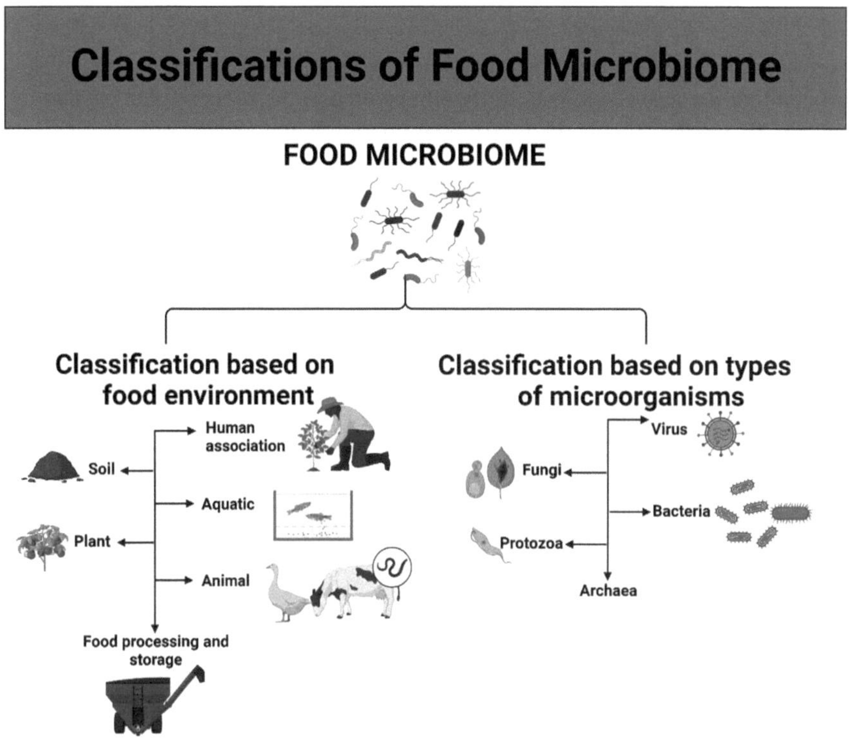

Fig. 6.1 Classifications of food microbiome (Image was created with Biorender.com)

cause black fungus in human with symptoms including runny nose, headache, fever, swollen and bulgy eyes, difficulty breathing and tissue necrosis (Suganya et al., 2019). Some species of *Fusarium* can also produce mycotoxins, which are detrimental to humans and animals (Qu et al., 2024).

iii. **Food Processing and Storage**

Food Processing is a process whereby food substances are converted to another form to increase consumption, appearance, and shelf life. This can be carried out in various processes involving chemical, mechanical, and physical conversion of raw food, which may reduce natural food microbiome biodiversity. Food contamination has also been recognized in association with food processing and storage activities, for example, bacteria like *Lactobacillus spp* and *Alicybacillus spp* are detected in food processing plants (Kumar et al., 2021; Patel et al., 2022a, 2022b).

iv. **Aquatic**

Food substances that originate from aquatic environments including different kinds of animals like crabs, mollusks and fish and plants like algae can harbor disease-causing pathogens and pose a significant effect on food safety (Sehnal et al., 2021).

v. **Plant**

Like soil microbiome, microbiome that interact with plants play a significant role in nitrogen fixation in the soil, especially those found in the roots of plants (Mahmud et al., 2020). On the other hand, food contamination with foodborne disease-causing microorganisms has been recognized in foods from plant sources. Plant-borne pathogens like *Phylophthora spp* (Li et al., 2014), *Erwimias amylovora* (Piqué et al., 2015) can cause gastrointestinal infections in animals when contaminated plant parts are ingested.

vi. **Animal**

Common disease-causing microbiome in foods from animal sources such as milk, eggs, and raw meat are *E. coli*, *Salmonella spp*, *Shigella* and *Salmonella spp* (Madilo et al., 2023).

6.2.2 Classification Based on Type of Microorganisms

Concerning food safety, we can classify microorganisms into five groups, namely (Lorenzo et al., 2018).

1. Bacteria
2. Virus
3. Archaea
4. Protozoa
5. Fungi.

This microbiome are important in food spoilage, foodborne pathogens, bioprocessing, and biorefinery.

6.3 History and Trends of Microbiome-Related Food Safety

Microorganisms have been used in food production since prehistoric times, and their function in fermentation processes was recognized before the development of bacteriology (Gokulrajan & Akash, 2023). Early humans were initially carnivorous, later including plants in their diet, and began cooking food (Jay et al., 2005). Around 6000 BC, cooking and improper food storage gave rise to food deterioration and poisoning (Tanner, 1994). Pottery-making, introduced to Western Europe from the Near East around 5000 BC, and the invention of boiler pots in the Near East about 8,000 years ago, facilitated cereal cooking, brewing, and food storage and subsequently beer production started in ancient Babylonia around 7000 BC (Stewart & Amerine, 2012).

The discovery of microorganisms in the 1830s led food producers to focus on optimizing their use. Over the next century, advancements such as genetic modification and DNA discovery enabled the creation of various foods, including beer, yogurt, cheese, soy sauce, wine, and vinegar (Hanlon & Sewalt, 2021). The top three foodborne pathogens in fresh foods were recognized by the Centers for Disease Control and Prevention (CDC) and the National Outbreak Reporting System (NORS) between 2008 and 2017, they include *Salmonella enterica, Escherichia coli and Norovirus* (Garcia et al., 2020). Among others, norovirus is the primary cause of foodborne diseases, which is responsible for 45% of cases associated with the consumption of contaminated fruits and vegetables (Gould et al., 2013; Yang et al., 2018). Analyzing food microbiome provides valuable insights into food quality by detecting spoilage-associated microorganisms. Microbial activity is the primary cause of food spoilage, leading to undesirable changes in sensory characteristics such as color, texture, and the development of odors and flavors (Jarvis et al., 2018). Foodborne disease is a pressing health concern in the United States. According to the Centres for Disease Control and Prevention (CDC), one in every six people (48 million individuals) contracts foodborne disease each year (Hoffmann et al., 2021). While most cases are mild, about 128,000 people require hospitalization each year. Serious long-term effects can include mental retardation, chronic gastrointestinal disorders, blindness, kidney failure, temporary paralysis, and reactive arthritis. Annually, approximately 3,000 individuals die from foodborne infections (Hoffmann et al., 2021). The United States food regulations mandate microbial testing for production facilities and high-risk food items, resulting in a strong reliance on microbial analysis to ensure food safety (Nayak & Waterson, 2019).

6.4 Challenges of Food Safety as a Result of Microbiome Interactions

The complexity of different microorganisms and their impact on shelf life, food quality, and the risk of foodborne illnesses pose several challenges on food safety (Sergaki et al., 2018). The challenges can be explained as follows:

6.4.1 Challenges of Foodborne Pathogens on Food Safety

Foodborne pathogens, including bacteria, fungi, viruses, and parasites, reduce food quality and pose a substantial control and maintenance challenge on food industries in maintaining food safety (Lourenco et al., 2022). Fungi such as *Aspergillus* and *Penicillium* can cause spoilage and toxicity in high-moisture environments. These fungi are ubiquitous and can proliferate and degrade agricultural products, especially in moist states. These degradations can be seen in colors, flavors and general products quality (Tarasova et al., 2022). They can also produce mycotoxins such as ochratoxin, which can contaminate food. This toxic secondary metabolite are often times carcinogenic, mutagenic, and immunosuppressant in nature (Choudhary et al., 2023). Foodborne bacteria pathogens can significantly reduce food safety through contamination. Viruses such as *Norovirus* and *Rotavirus* are infectious and can be transmitted through fecal oral route, infected individuals, and contaminated water (Agnihothram et al., 2016). When food are not cooked properly, it can cause transmission of these viruses and can result in vomiting, nausea, abdominal pain, and diarrhea, especially in children (Kluth et al., 2021). *Toxoplasma gondii* and *Cyclospora* can contaminate food through undercooked meat, unpasteurized dairy products, and contaminated water (Chlumsky et al., 2021). Biofilm formation by pathogens on food surfaces increases the risk of contamination and resistance to sanitizers while also protecting bacteria from the host immune system, allowing them to persist and cause food infections (Chlumsky et al., 2021). Biofilms are layers of microorganisms that grow on wet surfaces of food materials. Microorganisms also form biofilms on food processing and packaging equipment thus contaminating food and causing foodborne diseases (Zhao et al., 2017). These biofilms are capable of contaminating food through the toxins they produce and they are resistant to sanitization or cleaning, making them very dangerous (Elafify et al., 2024). For example, *Staphylococcus aureus* in biofilms have been reported to produce enterotoxins which caused food poisoning (Lianou et al. 2020). Fungi can also produce mycotoxins, which can be harmful to humans as well (Bintsis, 2017). *Bacillus cereus* a bacteria form biofilm and produce emetic and diarrhea toxins that can cause gastrointestinal problems (Chitlapilly Dass & Wang, 2022).

6.4.2 Challenges of Antimicrobial Resistance on Food Safety

Antimicrobial resistance is the process through which microbes develop the ability to survive exposure to an effective antimicrobial agent through mutation or gene transfer (Mancuso et al., 2021). This is a prominent issue in food safety, as it helps to increasing prevalence of resistant pathogens like *Salmonella* and *Campylobacter,* which are unaffected by antibiotics, and *Listeria monocytogenes*, which can form biofilms that protect bacteria from antimicrobial treatments. This leads to reduced treatment options, causing severe and prolonged food infection, and increased risks of poisoning (Silva et al., 2024). The presence of resistant bacteria can lead to the trying of several antibiotics on a particular bacteria and this can cause increase in the presence of antibiotics which can contamination soil and water and disrupt the ecosystems by harming wildlife (Jia et al., 2023). The rise in the need for antibiotics due to resistance can lead to soil, water, and air pollution. For water, this can cause poisoning for both aquatic organisms, other animals and humans. For soil pollution, this can prevent the growth of some florals or accumulation of chemicals in them resulting in abnormal human health and high hospitalization rate, high mortality rate if humans consume these plants. This can significantly affect the public health negatively (Ibrahim Mohamed, 2021). In low- and middle-income countries, ensuring food safety is particularly challenging due to weak regulations, inconsistent enforcement, and limited resources; these escalate antimicrobial resistance and affect food safety (Kasimanickam et al. 2021). In most cases, combating the issue of antimicrobial resistance requires the use of high quantity of antimicrobials and as such, this can lead to increase in greenhouse gas emission (Ahmed et al., 2024). Several antibiotics can also be tried to mitigate the resistance, and this can lead to damage to the ecosystem. The antibiotics could be harmful to farm animals and could be washed in water bodies. It could also result in the generation of several wastes from the antibiotics, which could lead to improper management and further increase of carbon emission (Chinemerem Nwobodo et al., 2022).

6.4.3 Challenges of Climate Change on Food Safety

Climate change can impede food safety in various ways. When there are changes in the pattern of rainfall and weather, this can lead to water scarcity, and as a result affect agricultural practices and food production (Devi, 2021; Tripathi, 2021). Food shortage and food insecurity can result from population growth (Eruaga, 2024). Part of the objective of the United Nations is to ensure food security and food safety, that was the reason why the sustainable development Goal 2 was created, which is aimed at ending hunger, achieving food security and improving nutrition and promoting sustainable agriculture (Montagnini & Metzel, 2024). Climate change such as rising temperature, extreme weather condition, bad precipitation, droughts, storms, and floods can prevent good agricultural practices and reduce crop yield, animal production thus causing shortage of food, food insecurity, reduction in agricultural workers income, loss of employment, and increased malnutrition (Tabari, 2020; Tripathi, 2021).

6.4.4 Government Policy and Regulations on Food Safety

Ensuring food safety is a common goal shared by industry and government. Each must carry out their distinct complementary roles to accomplish this shared objective. Ensuring food safety is primarily the responsibility of industry. It is the duty of the government to ensure that industries are carrying out their obligations (Tompkin, 2001). Foodborne diseases caused by microorganisms remain a significant issue, with these microbes entering the food chain at various points and producing toxic compounds. Though the food industries set up their quality control and assurance team to ensure food safety, it is the responsibility of the government and the regulatory bodies to ensure adequate compliance to the stipulated standards and to ensure good manufacturing practices (GMP). In this regard, a science-based approach can be employed to focus on predicting potential problems using risk analysis (Havelaar et al., 2010). Monitoring contamination in the food chain and tracking human disease outbreaks is crucial, with new methods such as molecular markers for outbreak detection, sero-epidemiology, and disease burden estimation improving surveillance (Wu et al., 2021). Advancements in molecular techniques will surely allow rapid development of new tracking methods and understanding microbial behavior under stress (Bari & Zaman, 2021).

Collaborating with the private sector, governments and their agencies may invest in genetically engineered organisms, which can improve crop yield, ensuring food safety and revolutionizing agriculture by enhancing crop traits, creating crop resistance to pest, diseases, and unpleasant climatic and environmental conditions through transgenic modification and genome editing (Yogi et al., 2024). Regarding this modification, care must be taken to prevent allergenicity or other consequences that may arise.

The government and other food agencies must also formulate and enforce rules guiding the production, processing, packaging, trading, and distribution of food. This will ensure promotion of food security and safety. Care must also be taken to ensure that the rules, which could later emerge as a global food law, do not contradict each other (Van der Meulen, 2010). The food law should also deal with issues of countries with many natural resources like land and other good agro conditions such as environmental conditions that can support agriculture but still rely on importation of food. If this is done, food security and safety will be maximized (Aginam, 2002). It is also imperative for governments to measure the correlations between food safety and climate change and to advance scientific understanding beyond empirical data (Smyth et al., 2021).

6.4.5 Challenges of Food Packaging on Food Safety

Food packaging has been helpful in protecting food from physical, chemical, and microbial contaminants. However, it can also cause spoilage due to contaminated packaging material, mishandling, or contact with non-sterile environments (Karanth et al., 2023). For example, the 1999 Salmonella outbreak in the United States was because of unhygienic conditions in orange juice mixing and bottling (Karanth et al., 2023). To prevent deterioration and guarantee food safety, food must be maintained at a temperature that will discourage the growth of microorganisms once it has been packaged. Environmental cleanliness, temperature, and humidity are a few of the critical factors that affect the safety and quality of food after packaging. Increased humidity can encourage the growth of microorganisms, excessive heat can lead to food spoilage, and inadequate hygiene can attract pests, which can disperse diseases (Alamri et al., 2021). However, the essence of packaging is for food safety and sustainability, sometimes, sustainability and food safety conflict (Santi et al., 2022). Packaging, for instance, can aid in preventing dangerous bacteria from growing on foods with a high moisture content, but it may also raise questions regarding the safety of food in general. For example, food safety regulations frequently support single-use packaging to prevent contamination, even though reusing containers is better for the environment (Vasile & Baican, 2021). Reusing packaging materials can make the supply chain more sustainable, but their lifecycle does not always fit well with a circular economy approach, which aims to minimize waste and make the most of resources. In this regard, regulatory bodies should encourage the adoption of sustainable and alternative packaging options to strike a balance between food safety and sustainability (Coelho et al., 2020).

6.4.5.1 Food Preservation

Food preservation techniques encompass a range of strategies used to prevent spoilage of food following harvesting or slaughtering. Preservation practices such as drying, refrigerating, and fermentation have been used for many years and have been complemented by other methods like canning, pasteurizing and the use of chemical preservatives. All these techniques are designed to prolong the shelf life of food products to ensure quality retention. Several progresses have been made in recent times to improve the quality of produce following the advancement in technology and knowledge about the causes (Gram et al., 2002). Despite the sophistication of these methods, they have faced several challenges that result in food spoilage. This spoilage remains a major concern for food industries worldwide, as it renders food unfit for human consumption, impacting both commercially produced and home-preserved items.

Microbial activities, including those of bacteria, yeast, and fungi pose a significant threat to the preservation of food products worldwide. These microorganisms may grow directly on the food or act on some microbial enzymes present in the food. The existence of these enzymes in food are usually released from the plants and animal tissues due to some human activities during and after harvest. Food quality degradation such as depreciation in texture, loss of nutrients, and change of taste are caused by chemical reactions due to enzymatic reactions. These actions will in return lead to texture changes in the produce, reduce nutritional value, and negatively affect the safety of fresh produce (Ray et al., 2005).

6.4.5.2 Harmful Microorganisms

The major example of pathogenic microorganisms that pose a significant threat to food safety are bacteria, yeast, and fungi. Bacteria are among the most common and dangerous foodborne pathogens. Notable examples include *Salmonella*, *Escherichia coli*, *Toxoplasma gondii,* and *Listeria monocytogenes*. These bacteria can contaminate food through various means, including poor hygiene practices, cross-contamination, and improper cooking or storage (Scallan et al., 2011). Research indicates that approximately more than 55,961 hospital cases and over 1300 deaths in the United States are attributable to foodborne illnesses resulting from microbial contamination. Norovirus accounts for about 58% of these illnesses, 11% is attributed to nontyphoidal *Salmonella* species. *Clostridium perfringens* makes up 10% and *Campylobacter species* at 9%. The primary contributors to hospitalizations are nontyphoidal *Salmonella species*, which contribute about 35%. Regarding fatalities, the leading causes are nontyphoidal *Salmonella* species, which leads to 28% of deaths, while *Listeria monocytogenes* and norovirus stand at 19 and 11%, respectively (Scallan et al., 2011).

Viruses, such as norovirus and hepatitis A, are also significant contributors to foodborne illnesses. Unlike bacteria, viruses do not grow on food but can be transmitted via contaminated food or water. Infected food handlers are often the source of viral contamination (Koopmans & Duizer, 2004). Noroviruses are classified under the family *Caliciviridae*, which comprises multiple genera. Among these, Norovirus and Sapovirus are harmful to humans. Norovirus have also been identified in various animals such as mice, pig, cattle, dogs, sheep, and cats, whereas sapoviruses have been found in pigs (Han et al., 2004; Martella et al., 2007). In humans, Norovirus primarily causes gastroenteritis while in animals. (Farkas et al., 2008). Currently, Norovirus and sapoviruses are recognized to cause diseases in humans when injected through food contamination (Scallan et al., 2011).

Fungi, including molds and yeasts, pose significant threats to food safety through spoilage and the production of mycotoxins, which are harmful to human health. Yeast and molds flourish in acidic environments (pH below 7). Specifically, yeasts have an optimal growth range of pH 3.5–4.5, while molds can thrive within the 3.5–8.0 pH range. Although fruits inhibit the growth of bacteria, due to their naturally low pH, it still supports the proliferation of yeasts and molds, leading to fruit spoilage (Singh & Desrosier, 2024).

Aflatoxins, produced by *Aspergillus* species, are particularly concerning due to their carcinogenic properties (Bennett & Klich, 2003). Research on fungal pathogens, particularly molds, highlights the critical importance of proper storage conditions to prevent fungal growth and mycotoxin production. Studies on aflatoxins have spurred the development of biocontrol methods and the establishment of regulatory limits, significantly contributing to enhanced food safety (Probst et al., 2007).

Parasites such as *Toxoplasma gondii* and *Trichinella spiralis* can be transmitted through contaminated meat produce. These parasites can cause severe illnesses, particularly in immunocompromised individuals (Pawlowski et al., 2012).

Studies on parasitic pathogens emphasize the importance of cooking meat to safe temperatures and implementing effective meat inspection programs to prevent infections. It shows the importance of educating the public about safe methods of food handling to minimize the risk of parasitic infections. (Dubey, 2009; Murrell & Pozio, 2011).

6.4.5.3 Cost of Control

The cost of controlling foodborne pathogens is substantial, encompassing various direct and indirect expenses.

a. **Direct costs**: Direct costs include expenses related to testing, monitoring, and implementing food safety measures. For instance, regular microbiological testing of food products and the environment is essential to ensure compliance with safety standards. Additionally, Implementing Hazard Analysis and Critical Control Points (HACCP) systems necessitates substantial investment in both training and infrastructure.
b. **Indirect costs**: Indirect costs involve the economic impact of foodborne illnesses on public health and the economy. These costs include medical expenses, lost productivity, and legal liabilities. The burden on the healthcare system is considerable, with costs arising from hospitalization, treatment, and long-term health consequences (Scharff, 2012).

The economic burden of foodborne illnesses can be illustrated through case studies. For example, the 2011 *Listeria* outbreak linked to cantaloupes in the United States resulted in 33 deaths and an estimated economic impact of over $150 million due to medical costs, product recalls, and lost sales (Batz et al., 2012).

6.4.6 Technological Impact

Technological advancements have had a profound impact on food safety, bringing both benefits and challenges. On a positive note, advancements in detection technologies, like polymerase chain reaction (PCR) and next-generation sequencing, have greatly improved our capacity to swiftly and precisely identify and trace foodborne pathogens. These cutting-edge methods enable rapid responses to contamination incidents, thereby minimizing the spread and impact of foodborne illnesses (Jaykus et al., 2019).

Conversely, the increasing complexity of modern food systems introduces new risks. The over-dependence on antibiotics in agriculture has led to the development of antibiotic-resistant bacteria, which in turn is transferred to humans when foods are consumed. Moreover, the globalization of food trade amplifies the potential for widespread contamination events, as contaminated products can be distributed across multiple regions.

6.5 Increase in Risk of Contamination as a Result of Globalization

Globalization has transformed food systems, creating opportunities and challenges for food safety.

i. Supply Chain Complexity

Global supply chains are inherently complex, involving multiple stages from production to consumption. This complexity increases the potential for contamination at various points in the supply chain. For example, raw materials sourced from different countries may have varying levels of safety standards and practices, leading to inconsistent quality and safety (Henson & Reardon, 2005).

ii. Rapid distribution

The rapid distribution of food products globally can facilitate the spread of contaminants. A contamination event in one part of the world can quickly become a global issue, as seen in the 2008 melamine contamination scandal in China, which affected dairy products worldwide (Pei et al., 2011).

iii. Regulatory agencies

Regulatory agencies face significant challenges in managing food safety in a globalized market. Harmonizing food safety standards across countries is difficult, and there is often a lack of consistent enforcement. This inconsistency can lead to gaps in safety measures and increased risk of contamination (Garcia Martinez et al., 2006).

6.6 History of Innovations in Microbiome Interactions in Food Safety

Food safety is a very relevant aspect of public health that has greatly developed over the years (Patarata et al., 2024). Food microbiology has developed all aspects, from DNA sequencing to metagenomics, probiotics, and biopreservation, in order to enhance food quality and safety (Høiby, 2024). The Germ theory of disease by Louis Pasteur and Robert Koch transformed the use of microorganisms in health and food safety (da Silva et al., 2023). Pasteurization is a method of killing bacteria that cause illness in food through heat application with the ultimate goal of reducing cases of foodborne diseases (Azizi-Lalabadi et al., 2023). In the late nineteenth and early twentieth centuries, the general adoption of refrigeration and freezing gave new ways of preventing microbial growth. Predictive microbiology represents an application of mathematical models that predict microorganism behavior in varying conditions. This data from microbiome feeds into quantitative risk assessment, which estimates the probability and consequences of microbial hazards (Graham & Ledesma-Amaro, 2023). State-of-the-art developments in synthetic biology further enabled the engineering of specific characteristic microbiome for food safety applications. Rapid detection methods and next-generation sequencing platforms will allow detailed profiling of the food microbiome to establish possible risks and implement targeted interventions (Elbehiry et al., 2023). Genomics technologies have impacted current microbiome research by making information on microbial strains available and enhancing food safety through competitive exclusion of harmful pathogens (da Silva et al., 2024). These heralds welcome an age for the development of enhanced safety and more nutritious foodstuffs.

In addition, food safety will continue to harness the power of biotechnology and data sciences in recalibrating engineered probiotics, investigating microbiome transplantation, synthetic biology, and other strategies to improve food safety, resilience of food systems, and sustainability while reducing reliance on chemical preservatives (Tay et al., 2023).

6.6.1 Examples

It has been highly recognized that microbiome interactions play a vital role in ensuring food safety. Advances in this regard can transform the approach to foodborne illness prevention, detection, and management. Research on the microbiome has advanced food safety by offering innovative approaches to enhance the quality, safety, and sustainability of food production. Other approaches to ensuring safe food, such as predictive microbiology, use mathematical modeling to anticipate pathogen growth and continue to develop improved preservation techniques to prevent foodborne illness. The innovative methods are discussed below (Evangelista et al., 2023; Taiwo et al., 2024).

6.6.1.1 Fermentation and Functional Foods

Fermentation and functional foods intersect traditional practices with modern nutritional science. Fermentation is a metabolic process by which sugars are selectively transformed into organic acids, gas, or alcohols using microorganisms such as bacteria, yeast, and fungi (Sun et al., 2022). Examples of common fermented foods are yogurt, kefir, sauerkraut, kimchi, miso, and soy products (tempeh and natto) (Akhter et al., 2024; Farooqui, 2021). Functional foods, however, have a potentially salutary effect on health beyond just providing nutrition. They are used to fortify other nutrients or bioactive elements with vitamins and minerals, and usually instrumental probiotics, for enhancing health and well-being (Peng et al., 2020). These foods contain probiotics and prebiotics in charge of gut health and bioactive compounds resulting from fermentation, which involves peptides, vitamins, and antioxidants (Aslam et al., 2020). These compounds lead to improved metabolic health, strengthened immunity, and reduced inflammation. The consumption of fermented functional foods can benefit digestive health by modulation of gut microbiota, enhancing immune function, and improving mental health (Chandrasekaran

Table 6.1 Examples of food item preserved by fermentation

Food item	Description	References
Kimchi	A traditional Korean dish made from fermented vegetables, primarily cabbage	Surya and Nugroho (2023)
Sauerkraut	Fermented cabbage that offers a tangy flavor and probiotics	Akhtar et al. (2021)
Kefir	A fermented milk drink similar to yogurt, rich in probiotics	Azizkhani et al. (2021)
Tempeh	Fermented soybeans forming a firm, nutty cake, rich in protein and probiotics	Ismail et al. (2022)
Miso	A Japanese seasoning produced by fermenting soybeans with salt and koji	Allwood et al. (2021)
Natto	Fermented soybeans known for their strong flavor and sticky texture	Liu et al. (2022)
Yogurt	Dairy products fermented with bacteria, providing probiotics	Kariyawasam et al. (2021)

et al., 2024). Probiotics help in maintaining a healthy gut lining through enhanced absorption of nutrients and preventing various gastrointestinal diseases such as diarrhea and irritable bowel syndrome (Cichońska et al., 2023). Fermented functional foods enhance the immune response, reducing the risk of infection. The composition of fermented foods makes them influential in brain function and mood through bioactive compounds in the foods, which work to reduce symptoms of anxiety and depression (Giuffrè & Giuffrè, 2024). In addition, fermentation technology sustains local economies through support for local semi-artisanal food production, not only helping to maintain the preservation of traditional foods but also to provide open markets for higher quality locally produced food. This artisanal nature can be linked directly to superiority in quality, as it contributes to their popularity and adds value to their economy (Khayeka-Wandabwa et al., 2024) (Fig. 6.1 and Table 6.1).

6.6.1.2 Pathogen Detection and Control

Pathogen detection and control innovations have huge impacts on microbiome interactions in food safety. New methods for the detection of pathogens, such as high-throughput sequencing and culturing on agar plates, permit the comprehensive examination of microbial communities (Hassoun et al., 2024). Improvements in the Non-Enzymatic Microbe Identification System have proved to be useful in *E. coli* detection and risk management with respect to pathogens in food (Shi et al., 2024). Modernization of control methods in agricultural and food-producing industries is focused on preventive strategies against pathogenic microorganisms to minimize the contamination of these products for food safety (Nath, 2024). Microbiome in food production chains are indicators that may show lurking hazards, underscoring the need to understand microbiome interactions for food safety management (Trollman et al.,

Table 6.2 Foods preserved by pathogen detection and control

Food item	Description	References
Dairy products	Includes milk, cheese, and yogurt, often tested for *Listeria* and *E. coli*	Rosario et al. (2021)
Meat products	Beef, pork, and poultry monitored for *Salmonella* and *E. coli* contamination	Williams et al. (2020)
Poultry	Chickens and turkeys tested for pathogens like *Campylobacter* and *Salmonella*	Tenhagen et al. (2021)
Seafood	Fish and shellfish are checked for *Vibrio* and *Listeria monocytogenes*	Dumen et al. (2020)
Fruits	Fresh and processed fruits tested for *E. coli* and *Salmonella*	Lee et al. (2022)
Vegetables	Leafy greens and other vegetables monitored for *Listeria* and *E. coli* contamination	Bolívar et al. (2023)
Eggs	Regularly tested for *Salmonella* to prevent outbreaks	Cardoso et al. (2021)
Processed foods	Includes ready-to-eat meals, often tested for multiple pathogens including *Listeria* and *Salmonella*	dos Santos et al. (2021)
Juices	Fresh and packaged juices checked for *E. coli* and *Salmonella*	Mendes-Oliveira et al. (2020)
Baked goods	Products like bread and pastries are monitored for mold and bacterial contamination	Paz-Arteaga et al. (2023)

2024). The integration of pathogen detection and control with preventive strategies leads to enhanced food safety by identifying and mitigating pathogen risks against the realization of safer production and consumption of foods (Upadhyay, 2020) (Table 6.2).

6.6.1.3 Prebiotics and Probiotics

The innovative developments in microbiome interactions, mainly concerning food safety, are prebiotics and probiotics (Mather et al., 2024). Prebiotics are non-digestible dietary fibers that work as a medium to stimulate the growth and specific activities of beneficial bacteria residing in the colon, thereby enhancing gut health and overall well-being (Yoo et al., 2024). Probiotics are living microbes that are introduced under controlled conditions to the host. They offer several advantages to health by colonization in the gut and rivaling pathogenic bacteria for survival; this enhances the balance and resilience of the gut microbiome (Melara et al., 2022). This interaction changes the gut microbiome to enterotype balance and diversity required for proper digestive and immune functions (Amenyogbe, 2023). The use of prebiotics and probiotics in foodstuffs has implications on food safety because some of these probiotics can inhibit the growth of pathogenic bacteria through competitive inhibition and the production of antimicrobial substances (Gomaa, 2020). *Lactobacillus* species, commonly found in fermented foods such as yogurt and cheese, have been shown to be effective in preventing the growth of harmful bacteria, such as *Salmonella* and *Listeria.* These beneficial bacteria employ various mechanisms to achieve this result, including the production of antimicrobial compounds such as bacteriocins, competition for nutrients, and modulation of the host immune system. *Lactobacillus plantarum* is known for its ability to produce antimicrobial bacteriocins. These bacteriocins have been shown to be effective in controlling the growth of *Listeria monocytogenes*, a pathogen in fermented foods (Ibrahim et al., 2021). Rabetafika et al. (2023) reported that *Lactobacillus acidophilus* also produces organic acids, such as lactic acid, which are utilized against the food pathogen *E. coli* O157 in dairy products. *Bifidobacterium bifidum* produces hydrogen peroxide, an antimicrobial agent used to control food pathogens such as yogurt and other dairy products (Rabetafika et al., 2023). *Pediococcus acidilactici* produces an antimicrobial substance known as pediocin that controls food pathogens in meat products, such as *Listeria monocytogenes* (Fijan, 2016). *Lactococcus lactis* also produces an antimicrobial agent known as Nisin, which is used against food pathogens such as *Clostridium botulinum* in processed cheeses (Chugh & Kamal-Eldin, 2020). *Saccharomyces cerevisiae* synthesizes antimicrobial agents, such as ethanol and organic acids, to control food pathogens, such as *Salmonella spp.*, in fermented beverages and bread (Chugh & Kamal-Eldin, 2020). They also prevent digestive disorders and protect against foodborne pathogens. Nondairy-derived prebiotics have been shown to reduce pathogenic populations in the gut, modulate total gut flora, and reduce the potential for foodborne illness (Mousavi Khaneghah et al., 2020). In poultry and livestock farming, competitive exclusion products introduce beneficial microbial communities into the gut of animals to reduce colonization by harmful pathogens, such as *E. coli* O157. The addition of a cocktail of benign bacteria to chicken diets can drastically reduce the prevalence of *Salmonella* in poultry products, thereby improving food safety from farm to table. However, there are regulatory considerations in adding prebiotics and probiotics to food products, as this is overseen by regulators who govern the labeling, formulation (Koirala & Anal, 2021), and health claims related to such novel ingredients to ensure consumer safety and ensure that consumers have the ability to make informed choices (De Carvalho et al., 2022).

6.6.1.4 Advanced Microbiome Monitoring and Metagenomics

Microbiome interactions, including advanced microbiome monitoring through metagenomics, have immense impacts on food safety (Farag et al., 2023). Advanced microbiome monitoring techniques use RNA to analyze microbial communities in food and thus power early hazard detection for both the quality and safety control of food (Ripolles-Avila et al., 2020). Genetic material originating from environmental samples enables research into the diversity and functional potential of microbial communities (Kumar et al., 2021). These techniques have facilitated pathogen, beneficial microbes, and spoiler detection in foodstuffs, and can be used for risk assessment and management associated with them (Banerjee et al., 2022). This new technology shifts the understanding of microbiome within food environments by providing tools for real-time monitoring, the ability to detect contaminants rapidly, and source tracking of microbes (Das et al., 2024). Metagenomic surveillance of foodborne pathogens has been used for metagenomic microbial source tracking in the food industry and regulatory sectors. This approach has facilitated tracing of foodborne pathogens and fecal contamination back to their sources. Consequently, food safety and regulatory measures have been enhanced (Ahlinder et al., 2022). It has also been used to enhance food safety by employing metabarcoding and source tracking to detect microorganisms during food fermentation. This method was applied to identify the sources of both desirable and spoilage microorganisms in black olive fermentation, leading to improved management of the fermentation processes and overall food safety (Billington et al., 2022).

Omic-based monitoring of microbial dynamics was performed to investigate microbial dynamics across the food chain. Metagenome-assembled genomes (MAGs) of relevant species were retrieved, which have proven useful for strain tracking along the food chain and in instances of food poisoning. This approach has improved our understanding of microbial dynamics and safety in the food supply chain (Sequino et al., 2022).

The development of a microbiome analysis pipeline for food safety and quality monitoring has been achieved through the application of microbiome analysis for Food Safety. This innovative approach results in an elevated level of specificity for detecting microbes in food matrices, which is crucial for ensuring the safety and quality of food products (Beck et al., 2021). The precise identification of microorganisms enables the effective monitoring and assurance of food safety and quality. They form the foundation for preventing foodborne illnesses by enabling interventions that target the source and enhance the overall quality of food. Food safety strategies, precision agriculture, and tailored interventions based on specific microbial profiles show potential for the future (Galanakis, 2024).

6.6.1.5 Biopreservation with Natural Antimicrobials

Biopreservation is a process of shelf-life extension and improvement of the safety of food with natural substances, principally antimicrobials emanating from plants, animals, bacteria, or fungi (Aguirre-Garcia et al., 2024). This involves antimicrobials that can be produced from plant extracts, essential oils, organic acids, and bacteriocins originating from lactic acid bacteria, which inhibit pathogenic microorganisms and enhance the growth of valuable microbes (Garin-Murguialday et al., 2024; Xu et al., 2024). This selective control allows for the maintenance of a balanced microbial ecosystem and helps improve food safety and quality. Biopreservation is an innovative strategy in food safety to reduce dependence on synthetic preservatives and chemical additives, thus meeting consumer preferences for natural and clean-label products (Bodie et al., 2024). Biopreservation effectively addresses challenges related to microbial contamination and food deterioration during production and storage. This can be achieved by extending the shelf life of food products by controlling microbial activity, reducing food spoilage, and minimizing the likelihood of foodborne illnesses. Moreover, it supports clean label trends by utilizing natural ingredients, thereby catering to consumer demand for safer and healthier food options. Furthermore, biopreservation is effective across a wide range of food matrices, from meat to dairy products, and it addresses specific pathogens and spoilage microorganisms (Rhouma et al., 2024; Snyder et al., 2024). This approach represents a significant step towards enhancing food safety and quality.

6.6.1.6 Biosensing Techniques for the Detection of Pathogens

Traditional methods of detecting pathogens can be time-consuming, expensive, and require high technical know-how. Biosensors have gained a lot of attention over the years due to their ability to provide rapid, real-time results without compromising sensitivity.

A biosensor comprises main two components: a bioactive substance and a transducer (Naresh & Lee, 2021). The bioactive component can be enzymes, antibodies, or microorganisms that selectively recognize specific signals. These signals are converted into measurable output by the transducer (Chadha et al., 2022). Biosensors can be classified based on either the bioreceptor or the signal transducer (Fig. 6.2).

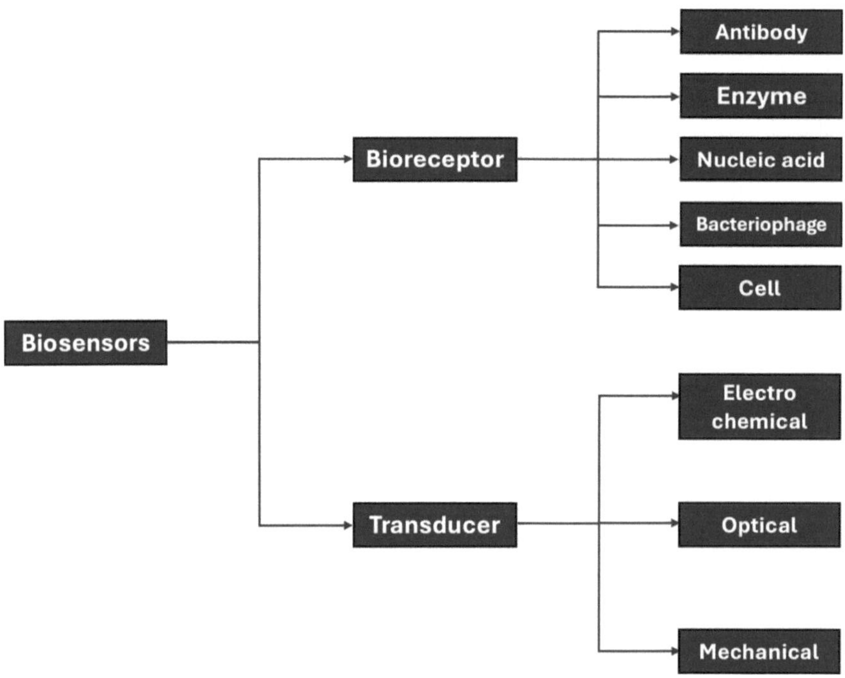

Fig. 6.2 Classification of biosensors (Adapted from Chadha et al., 2022; Naresh & Lee, 2021)

Classification based on bioreceptors includes antibody sensors, enzyme sensors, nucleic acid sensors, or bacteriophage, cell sensors. Antibodies are Y-shaped immunoglobulins composed of two identical pairs of heavy and light polypeptide chains held together by disulfide bonds (Pantaleo et al., 2022). They bind antigens via specific reception sites, making them very selective, and are capable of interacting with target molecules present in minute quantities (Nesakumar et al., 2021). These antibodies are usually deposited on the surface of the electrode in electrochemical sensors where they are commonly utilized (Cesewski & Johnson, 2020). Biocatalyst bioreceptors include enzymes that have intrinsic catalytic properties. These proteins speed up the rate of reactions and possess active binding sites that the pathogen substrates can be attached to. This intrinsic property allows high sensitivity towards pathogens/bacteria.

Transducers could be electrochemical, optical, or mechanical depending on the signals being analyzed (Chen et al., 2018). The electrochemical transducer utilizes electrodes to convert chemical energy, from the interactions with recognition components, into electrical energy (Cesewski & Johnson, 2020). Optical transducers use light signals resulting from changes in optical properties after analyte/bioreceptor interaction. A few include fluorescence, reflectance, absorbance, and surface plasmon resonance (SPR)-based transducers (Chen et al., 2018). Mechanical transducers also known as piezoelectric function in response to changes in mechanical stress to generate electrical signals.

Faster pathogen detection time, high sensitivity and selectivity, and low cost of biosensors have made it applicable in a wide variety of fields like agriculture, food and water quality control, health care, and security. In food safety, accurate detection of pathogens (such as *E.coli, Salmonella typhimurium*), mycotoxins, and pesticides is important and can be achieved using biosensors (Griesche & Baeumner, 2020). Agriculture remains the major source of food and has been greatly impacted by the use of biosensors. These devices can be used for the detection of pathogens like aflatoxin, which causes liver damage, before or after crop harvesting. Hence, biosensors offer on-site, non-destructive, and rapid assessment of food product quality (Chadha et al., 2022).

For the detection of aflatoxins, biosensors incorporating metal nanomaterials have been developed. These are biocompatible and possess unique optical properties that are utilized in the detection application. Seyed et al. used gold nanoparticles (AuNPs) in the synthesis of aptasensor M1 which are very sensitive and selective (Jalalian et al., 2018). As an alternative to metal nanoparticles, metal oxides have been employed due to their relatively lower cost and higher stability. An antibody biosensor designed by Singh et al. (2013), utilized Sm2O3 nanorods and an electrochemical transducer for the detection of AFB1 in food, achieving a detection limit of 10–700 pg/mL. Carbon nanomaterials, like carbon nanotubes/fibers, quantum dots, and fullerenes, have been used due to their strong adsorption properties and mechanical controllability (Yan et al., 2020). Pesticides such as carbamates and organophosphates in food and crops have been detected using various electrochemical and optical biosensors. These sensors, enhanced with metal, carbon, or hybrid nanomaterials, improve efficiency, accuracy, and cost-effectiveness (Mirres et al., 2022).

6.6.1.7 Nanotechnology Innovations in Food Packaging

In ensuring food safety, proper food packaging is essential to preserve the quality of food, prevent contamination, and prolong its shelf life. Many innovations in the field of nanotechnology have been developed to ensure this goal is met while ensuring the eco-friendly nature of these materials. Beyond food packaging, nanotechnology finds application in improving crop yield and for environment monitoring which are also key aspects to ensure food safety.

Nanotechnology involves the creation, characterization, and modification of materials on a nanoscale, typically 1–100 nm. Due to their small size, nanomaterials offer a high surface-to-volume ratio and enhanced properties. Nanomaterials in food packaging do not imply that the final packaging product is on a nanoscale; rather about 50% of its composing particles are within the nano range (Enescu et al., 2019). The significant rise of research in this area over the years is due to the possibility of prolonging food shelf life and quality by effectively preventing the passage of moisture or gasses like CO_2 and O_2, which could allow microbial activity making food unsafe.

Packaging nanomaterials are classified into smart and active packaging. Smart packaging incorporates nanosensors capable of detecting changes like temperature, moisture, and gasses. For example, visual monitoring of the modified atmosphere packages (MAPs) in packaged foods for CO_2 with silica nanoparticles and fluorophore dyes (Kishore et al., 2024). These smart packaging possess indicators that communicate to the consumer the state of the food instead of relying solely on the best-before dates which could be irrelevant if, for example, the cold chain has been broken. In active packaging, nanoparticles impregnated in the polymer matrix are used to allow the release or removal of substances into or from the packaged food. Some functions include antioxidant and antimicrobial activity. Oxygen scavengers used like iron-based scavengers, activated by light, heat, or moisture, are loaded into films or sachets to capture O_2 which could otherwise lower the quality of food. Different metals and metal oxides such as silver and copper metal ions, titanium oxide, gold oxide, and magnesium oxide have been utilized to ensure antimicrobial activity (Primožič et al., 2021).

There have been increasing concerns about the safety of nanomaterials used in packaging materials due to the potential transfer of materials to the food itself, which could pose toxicity risks, as well as issues related to their biodegradability. Further research is ongoing to address some of these issues and encourage public acceptance of advances in this field.

6.6.1.8 Bioreactor for Sustainable Food Packaging Production

Food safety is multifaceted and can be approached from different angles. Plants remain a key source of food, nutrients, and minerals for humans and animals. During harvesting, processing, and storage of food, a lot of waste is generated. With the aim of converting trash to treasure, the food packaging industry uses biorefineries to convert agricultural waste into biodegradable products (Cristofoli et al., 2023). In the conversion of biomass like banana peels, rice husk, and corn waste to food packaging biomaterials, bioreactors are essential.

Bioreactors are devices that provide a controlled environment for different processes like fermentation, cell culture, biochemical conversion, and anaerobic digestion. Biorefineries convert waste to eco-friendly products. They can be classified based on how stirring or agitation occurs; some include stirred tanks, airlift, nutrient mist, and temporary immersion bioreactors (Werner et al., 2018).

These packaging biomaterials are not only comparable in looks and use to synthetic materials but also have the advantage of mitigating concerns like toxicity and recyclability. Additionally, there have been reports on the use of bioreactors to produce microbial colorants useful for food packaging biomaterials (De Oliveira et al., 2020). This innovation can replace conventional synthetic coloring due to its nontoxic properties and its ability to protect against oxidative damage.

6.6.1.9 Innovation in Food Preservation

Age-long food preservation techniques such as drying, heating, and salting have been known and used over the years to prevent food spoilage and ensure food safety. Most of the food preservation techniques are achieved by controlling water activity, pH, atmosphere composition, and temperature (Jans et al., 2016). Innovations in food preservation take one of two forms: development of food packaging materials or using preservatives.

Innovations in food packaging designed to preserve food quality have been discussed in a previous section, so the focus here will shift to food preservatives. There has been an increase in the use of essential oils and plant extracts as preservatives due to the presence of compounds like phenolic compounds and terpenes which possess antimicrobial and antioxidant properties (Batiha et al., 2021). These extracts are believed to be non-hazardous and eco-friendly. Antimicrobial peptides (AMPs) and anti-fungal proteins such as bacteriocins and lactoferrin respectively, have also been used as food preservatives (Boukoufi et al., 2022). Their activity, which is hinged on the intrinsic cationic nature, could be affected by pH and salt concentrations, potentially restricting their application. Chemical approaches are being developed to synthesize AMPs similar to those naturally extracted from plants but void of the aforementioned limitations (Shwaiki et al., 2021).

6.7 Conclusion

Food safety is a critical issue, especially in developing and underdeveloped nations where foodborne illnesses are prevalent. Ensuring food safety requires regulating the complex interactions between food microbiome and pathogens, which necessitates extensive risk-mitigation methods. This chapter explored several critical factors influencing global food safety, such as antimicrobial resistance, climate change, government policies and regulations, as well as advancements in food packaging and preservation. To address these issues, a holistic strategy is required, which includes better detection tools, enhanced food preservation techniques, and strict government controls. Technological innovations like nanotechnology and genetically modified organisms offer promising solutions. Collaboration among the private, government, and scientific communities is critical for establishing successful strategies and implementing strong food safety systems. Future developments will involve the incorporation of cutting-edge technology such as artificial intelligence and machine learning, which will revolutionize monitoring and traceability systems. Sustainable and eco-friendly practices in food packaging and preservation will minimize environmental impact while ensuring food safety. In addition, international collaboration and harmonization of food safety standards will be essential for a unified and resilient food safety framework.

References

Agnihothram, S. S. et al. (2016). Silicon dioxide impedes antiviral response and causes genotoxic insult during calicivirus replication. *Journal of Nanoscience and Nanotechnology, 16*(7), 7720–7730. https://doi.org/10.1166/jnn.2016.12828

Aginam, O. (2002). International law and communicable diseases. *Bulletin of the World Health Organization, 80*(12), 946–951.

Aggarwal, N. et al. (2023). Microbiome and human health: Current understanding, engineering, and enabling technologies. *Chemical Reviews. American Chemical Society*, 31–72. https://doi.org/10.1021/acs.chemrev.2c00431

Aguirre-Garcia, Y. L., Nery-Flores, S. D., Campos-Muzquiz, L. G., Flores-Gallegos, A. C., Palomo-Ligas, L., Ascacio-Valdés, J. A., Sepúlveda-Torres, L., & Rodríguez-Herrera, R. (2024). Lactic acid fermentation in the food industry and bio-preservation of food. *Fermentation, 10*(3), 168. https://doi.org/10.3390/fermentation10030168

Ahlinder, J., Svedberg, A.-L., Nystedt, A., Dryselius, R., Jacobsson, K., Hägglund, M., Brindefalk, B., Forsman, M., Ottoson, J., & Troell, K. (2022). Use of metagenomic microbial source tracking to investigate the source of a foodborne outbreak of cryptosporidiosis. *Food and Waterborne Parasitology, 26*, Article e00142. https://doi.org/10.1016/j.fawpar.2021.e00142

Ahmed, S. K. et al. (2024). Antimicrobial resistance: Impacts, challenges, and future prospects. *Journal of Medicine, Surgery, and Public Health*, 2, 100081. https://doi.org/10.1016/j.glmedi.2024.100081

Akhter, N., Majid, D., Rather, J. A., Majeed, M., Mehvish, S. T., Dar, B. N., Makroo, H. A., Muzaffar, K., & Sofi, S. A. (2024). A comprehensive review on formulation of baby foods: Nutraceutical potential of fruits, vegetables, cereals and probiotics in baby food preparation. *Current Food Science and Technology Reports*. https://doi.org/10.1007/s43555-024-00034-8

Akhtar, G., Ahmad Bhat, N., Masoodi, F. A., & Gani, A. (2021). Small- and large-scale production of probiotic foods, probiotic potential and nutritional benefits. In *advances in probiotics* (pp. 365–395). Elsevier. https://doi.org/10.1016/B978-0-12-822909-5.00023-X

Alamri, M. S., et al. (2021). Food packaging's materials: A food safety perspective. *Saudi Journal of Biological Sciences, 28*(8), 4490–4499. https://doi.org/10.1016/j.sjbs.2021.04.047

Allwood, J. G., Wakeling, L. T., & Bean, D. C. (2021). Fermentation and the microbial community of Japanese koji and miso: A review. *Journal of Food Science, 86*(6), 2194–2207. https://doi.org/10.1111/1750-3841.15773

Amenyogbe, E. (2023). Application of probiotics for sustainable and environment-friendly aquaculture management—a review. *Cogent Food & Agriculture, 9*(1). https://doi.org/10.1080/23311932.2023.2226425

Aslam, H., Green, J., Jacka, F. N., Collier, F., Berk, M., Pasco, J., & Dawson, S. L. (2020). Fermented foods, the gut and mental health: A mechanistic overview with implications for depression and anxiety. *Nutritional Neuroscience, 23*(9), 659–671. https://doi.org/10.1080/1028415X.2018.1544332

Azizi-Lalabadi, M., Moghaddam, N. R., & Jafari, S. M. (2023). Pasteurization in the food industry. In *Thermal processing of food products by steam and hot water* (pp. 247–273). Elsevier. https://doi.org/10.1016/B978-0-12-818616-9.00009-2

Azizkhani, M., Saris, P. E. J., & Baniasadi, M. (2021). An in-vitro assessment of antifungal and antibacterial activity of cow, camel, ewe, and goat milk kefir and probiotic yogurt. *Journal of Food Measurement and Characterization, 15* (1), 406–415. https://doi.org/10.1007/s11694-020-00645-4

Banerjee, G., Agarwal, S., Marshall, A., Jones, D. H., Sulaiman, I. M., Sur, S., & Banerjee, P. (2022). Application of advanced genomic tools in food safety rapid diagnostics: Challenges and opportunities. *Current Opinion in Food Science, 47*, Article 100886. https://doi.org/10.1016/j.cofs.2022.100886

Bari, M. L., & Zaman, S. (2021) Microbial biotechnology in food and health: present and future food safety regulation. In *Microbial biotechnology in food and health* (pp. 1–20). Elsevier. https://doi.org/10.1016/B978-0-12-819813-1.00001-3

Batz, M. B., Hoffmann, S., & Morris, J. G. (2012). Estimating the economic burden of foodborne illnesses. *Journal of Food Protection, 75*(8), 1233–1241. https://doi.org/10.4315/0362-028X.JFP-11-417

Batiha, G. E. S., Hussein, D. E., Algammal, A. M., George, T. T., Jeandet, P., Al-Snafi, A. E., Tiwari, A., Pagnossa, J. P., Lima, C. M., Thorat, N. D., & Zahoor, M. (2021). Application of natural antimicrobials in food preservation: Recent views. *Food Control, 126*, Article 108066. https://doi.org/10.1016/j.foodcont.2021.108066

Beck, K. L., Haiminen, N., Chambliss, D., Edlund, S., Kunitomi, M., Huang, B. C., Kong, N., Ganesan, B., Baker, R., Markwell, P., Kawas, B., Davis, M., Prill, R. J., Krishnareddy, H., Seabolt, E., Marlowe, C. H., Pierre, S., Quintanar, A., Parida, L., & Weimer, B. C. et al. (2021). Monitoring the microbiome for food safety and quality using deep shotgun sequencing. *Npj Science of Food, 5*(1), 3. https://doi.org/10.1038/s41538-020-00083-y

Bennett, J. W. & Klich, M. (2003). Mycotoxins. *Clinical Microbiology Reviews, 16*(3), 497–516. https://doi.org/10.1128/CMR.16.3.497-516.2003

Billington, C., Kingsbury, J. M., & Rivas, L. (2022). Metagenomics approaches for improving food safety: A review. *Journal of Food Protection, 85*(3), 448–464. https://doi.org/10.4315/JFP-21-301

Bintsis, T. (2017). Foodborne pathogens. *AIMS Microbiology, 3*(3), 529–563. https://doi.org/10.3934/microbiol.2017.3.529

Blum, W. E. H., Zechmeister-Boltenstern, S., & Keiblinger, K. M. (2019). Does soil contribute to the human gut microbiome? *Microorganisms, 7*(9). https://doi.org/10.3390/microorganisms7090287

Bodie, A. R., Wythe, L. A., Dittoe, D. K., Rothrock, M. J., O'Bryan, C. A., & Ricke, S. C. (2024). Alternative additives for organic and natural ready-to-eat meats to control spoilage and maintain shelf life: Current perspectives in the United States. *Foods, 13*(3), 464. https://doi.org/10.3390/foods13030464

Bolívar, A., Saiz-Abajo, M. J., García-Gimeno, R. M., Petri-Ortega, E., Díez-Leturia, M., González, D., Vitas, A. I., & Pérez-Rodríguez, F. (2023). Cross contamination of *Escherichia coli* O157:H7 in fresh-cut leafy vegetables: Derivation of a food safety objective and other risk management metrics. *Food Control, 147*, 109599. https://doi.org/10.1016/j.foodcont.2022.109599

Boukoufi, C., Boudier, A., Maincent, P., Vigneron, J., & Clarot, I. (2022). Food-inspired innovations to improve the stability of active pharmaceutical ingredients. *International Journal of Pharmaceutics, 623*, Article 121881. https://doi.org/10.1016/j.ijpharm.2022.121881

Cardoso, M. J., Nicolau, A. I., Borda, D., Nielsen, L., Maia, R. L., Møretrø, T., Ferreira, V., Knøchel, S., Langsrud, S., & Teixeira, P. (2021). Salmonella in eggs: From shopping to consumption—A review providing an evidencebased analysis of risk factors. *Comprehensive Reviews in Food Science and Food Safety, 20*(3), 2716–2741. https://doi.org/10.1111/1541-4337.12753

Cesewski, E., & Johnson, B. N. (2020). Electrochemical biosensors for pathogen detection. *Biosensors and Bioelectronics, 159*, Article 112214. https://doi.org/10.1016/j.bios.2020.112214

Chadha, U., Bhardwaj, P., Agarwal, R., Rawat, P., Agarwal, R., Gupta, I., Panjwani, M., Singh, S., Ahuja, C., Selvaraj, S. K., & Banavoth, M. (2022). Recent progress and growth in biosensors technology: A critical review. *Journal of Industrial and Engineering Chemistry, 109*, 21–51. https://doi.org/10.1016/j.jiec.2022.02.010

Chandrasekaran, P., Weiskirchen, S., & Weiskirchen, R. (2024). Effects of probiotics on Gut Microbiota: An overview. *International Journal of Molecular Sciences, 25*(11), 6022. https://doi.org/10.3390/ijms25116022

Chen, Y., Wang, Z., Liu, Y., Wang, X., Li, Y., Ma, P., Gu, B., & Li, H. (2018). Recent advances in rapid pathogen detection method based on biosensors. *European Journal of Clinical Microbiology & Infectious Diseases, 37*, 1021–1037. https://doi.org/10.1007/s10096-018-3230-x

Chinemerem Nwobodo, D. et al. (2022). Antibiotic resistance: The challenges and some emerging strategies for tackling a global menace. *Journal of Clinical Laboratory Analysis, 36*(9). https://doi.org/10.1002/jcla.24655

Chitlapilly Dass, S., & Wang, R. (2022). Biofilm through the looking glass: A microbial food safety perspective. *Pathogens, 11*(3), 346. https://doi.org/10.3390/pathogens11030346

Chlumsky, O., et al. (2021). Evaluation of the Antimicrobial Efficacy of N-Acetyl-l-Cysteine, Rhamnolipids, and Usnic Acid—novel approaches to fight food-borne pathogens. *International Journal of Molecular Sciences, 22*(21), 11307. https://doi.org/10.3390/ijms222111307

Choudhary, K. G., Sharfuddin, C., & Kumar, A. (2023). Comparative quantification by HPLC-FLD of Ochratoxin production by Aspergillus ochraceous MTCC 4643. *Ecology, Environment and Conservation, 29*(04), 1761–1765. https://doi.org/10.53550/EEC.2023.v29i04.051

Chugh, B., & Kamal-Eldin, A. (2020). Bioactive compounds produced by probiotics in food products. *Current Opinion in Food Science, 32*, 76–82. https://doi.org/10.1016/j.cofs.2020.02.003

Cichońska, P., Kowalska, E., & Ziarno, M. (2023). The Survival of psychobiotics in fermented food and the gastrointestinal tract: a review. *Microorganisms, 11*(4), 996. https://doi.org/10.3390/microorganisms11040996

Coelho, P. M., et al. (2020). Sustainability of reusable packaging–Current situation and trends. *Resources, Conservation & Recycling: X, 6*, Article 100037. https://doi.org/10.1016/j.rcrx.2020.100037

Cristofoli, N. L., Lima, A. R., Tchonkouang, R. D., Quintino, A. C., & Vieira, M. C. (2023). Advances in the food packaging production from agri-food waste and by-products: Market trends for a sustainable development. *Sustainability, 15*(7), 6153. https://doi.org/10.3390/su15076153

Da Silva, A. K., Perim, M. D., Brito, L. M., & Alvarenga, V. O. (2023). *Basic concepts for predictive microbiology* (pp. 1–30). https://doi.org/10.1007/978-1-0716-3413-4_1

Da Silva, T. F., Glória, R. de A., Americo, M. F., Freitas, A. dos S., de Jesus, L. C. L., Barroso, F. A. L., Laguna, J. G., Coelho-Rocha, N. D., Tavares, L. M., le Loir, Y., Jan, G., Guédon, É., & Azevedo, V. A. de C. (2024). Unlocking the potential of probiotics: A comprehensive review on research, production, and regulation of probiotics. *Probiotics and antimicrobial proteins*. https://doi.org/10.1007/s12602-024-10247-x

Das, M., Albert, V., Das, S., Dolma, K. G., Majumdar, T., Baruah, P. J., Hazarika, S. C., Apum, B., & Ramamurthy, T. (2024). An integrated FoodNet in Northeast India: Fostering one health approach to fortify public health. *BMC Public Health, 24*(1), 451. https://doi.org/10.1186/s12889-024-18007-w

De Carvalho, N. M., Oliveira, D. L., Costa, C. M., Pintado, M., & Madureira, A. R. (2022). Can Supplemented Skim Milk (SKM) boost your gut health? *Fermentation, 8*(3), 126. https://doi.org/10.3390/fermentation8030126

De Oliveira, F., Lima, C. D. A., Lopes, A. M., Marques, D. D. A. V., Druzian, J. I., Pessoa Junior, A., & Santos-Ebinuma, V. C. (2020). Microbial colorants production in stirred-tank bioreactor and their incorporation in an alternative food packaging biomaterial. *Journal of Fungi, 6*(4), 264. https://doi.org/10.3390/jof6040264

Dekaboruah, E., et al. (2020). Human microbiome: An academic update on human body site specific surveillance and its possible role. *Archives of Microbiology, 202*(8), 2147–2167. https://doi.org/10.1007/s00203-020-01931-x

Devi, R. M. (2021). Climate change: breeding strategies to mitigate abiotic stress in ornamental crops. In *Climate dynamics in horticultural science* (Vol. 1, pp. 251–272). Apple Academic Press. https://doi.org/10.1201/b18035-19

dos Santos, J. S., Biduski, B., & dos Santos, L. R. (2021). Listeria monocytogenes: health risk and a challenge for food processing establishments. *Archives of Microbiology, 203*(10), 5907–5919. https://doi.org/10.1007/s00203-021-02590-2

Dubey, J. P. (2009). History of the discovery of the life cycle of Toxoplasma gondii. *International Journal for Parasitology, 39*(8), 877–882. https://doi.org/10.1016/j.ijpara.2009.01.005

Dumen, E., Ekici, G., Ergin, S., & Bayrakal, G. M. (2020). Presence of foodborne pathogens in seafood and risk ranking for pathogens. *Foodborne Pathogens and Disease, 17*(9), 541–546. https://doi.org/10.1089/fpd.2019.2753

Elafify, M. et al. (2024). Biofilm formation in food industries: Challenges and control strategies for food safety. *Food Research International, 190*, 114650. https://doi.org/10.1016/j.foodres.2024.114650

Elbehiry, A., Abalkhail, A., Marzouk, E., Elmanssury, A. E., Almuzaini, A. M., Alfheeaid, H., Alshahrani, M. T., Huraysh, N., Ibrahem, M., Alzaben, F., Alanazi, F., Alzaben, M., Anagreyyah, S. A., Bayameen, A. M., Draz, A., & Abu-Okail, A. (2023). An overview of the public health challenges in diagnosing and controlling human foodborne pathogens. *Vaccines, 11*(4), 725. https://doi.org/10.3390/vaccines11040725

Enescu, D., Cerqueira, M. A., Fucinos, P., & Pastrana, L. M. (2019). Recent advances and challenges on applications of nanotechnology in food packaging. A literature review. *Food and Chemical Toxicology, 134*, 110814. https://doi.org/10.1016/j.fct.2019.110814

Evangelista, A. G., Danielski, G. M., Corrêa, J. A. F., Cavalari, C. M. de A., Souza, I. R., Luciano, F. B., & Macedo, R. E. F. de. (2023). Carnobacterium as a bioprotective and potential probiotic culture to improve food quality, food safety, and human health–a scoping review. *Critical Reviews in Food Science and Nutrition, 63*(24), 6946–6959. https://doi.org/10.1080/10408398.2022.2038079

Farag, M. A., Baky, M. H., von Bergen, M., & Hegazi, N. M. (2023). The use of omics in monitoring food gut microbiota interaction outcomes: A review of novel trends and technologies. *Current Opinion in Food Science, 52*, Article 101064. https://doi.org/10.1016/j.cofs.2023.101064

Farkas, T., Sestak, K., Wei, C., & Jiang, X. (2008). Characterization of a rhesus monkey calicivirus representing a new genus of Caliciviridae. *Journal of Virology, 82*(11), 5408–5416. https://doi.org/10.1128/JVI.00070-08

Farooqui, A. A. (2021). Importance of fermented foods on human health. In *Gut Microbiota in Neurologic and visceral diseases* (pp. 69–86). Elsevier. https://doi.org/10.1016/B978-0-12-821039-0.00012-5

Fijan, S. (2016). Antimicrobial effect of probiotics against common pathogens. In *Probiotics and Prebiotics in Human Nutrition and Health*. InTech. https://doi.org/10.5772/63141

Galanakis, C. M. (2024). The future of food. *Foods, 13*(4), 506. https://doi.org/10.3390/foods13040506

Garcia Martinez, M., Fearne, A., & Caswell, J. A. (2006). Co-regulation as a possible model for food safety governance: Opportunities for public-private partnerships. *Food Policy, 31*(5), 484–512. https://doi.org/10.1016/j.foodpol.2006.06.010

Garcia, S. N., Osburn, B. I., & Jay-Russell, M. T. (2020). One health for food safety, food security, and sustainable food production. *Frontiers in Sustainable Food Systems, 4*(January), 1–9. https://doi.org/10.3389/fsufs.2020.00001

Garin-Murguialday, N., Espina, L., Virto, R., & Pagán, R. (2024). Lactic acid bacteria and Bacillus subtilis as potential protective cultures for biopreservation in the food industry. *Applied Sciences, 14*(10), 4016. https://doi.org/10.3390/app14104016

Giuffrè, D., & Giuffrè, A. M. (2024). Fermentation technology and functional foods. *Frontiers in Bioscience-Elite, 16*(1), 8. https://doi.org/10.31083/j.fbe1601008

Gokulrajan, S., & Akash, K. (2023). Role of microbes in fermented food and food products-a review. In *Encyclopedia of agriculture and allied sciences* (pp. 138–147).

Gomaa, E. Z. (2020). Human gut microbiota/microbiome in health and diseases: A review. *Antonie Van Leeuwenhoek, 113*(12), 2019–2040. https://doi.org/10.1007/s10482-020-01474-7

Gould, L. H., Walsh, K. A., Vieira, A. R., Herman, K., Williams, I. T., & Hall, A. J. (2013). Surveillance for foodborne disease outbreaks—United States, 1998–2008 (Vol. 62, pp. 1–34). https://www.cdc.Gov/Mmwr/Preview/Mmwrhtml/Ss6202a1.Htm

Graham, A. E., & Ledesma-Amaro, R. (2023). The microbial food revolution. *Nature Communications, 14*(1), 2231. https://doi.org/10.1038/s41467-023-37891-1

Gram, L., Ravn, L., Rasch, M., Bruhn, J. B., Christensen, A. B., & Givskov, M. (2002). Food spoilage—interactions between food spoilage bacteria. *International Journal of Food Microbiology, 78*(1–2), 79–97. https://doi.org/10.1016/s0168-1605(02)00233-7

Griesche, C., & Baeumner, A. J. (2020). *Biosensors to Support Sustainable Agriculture and Food Safety.* https://doi.org/10.1016/j.trac.2020.115906

Han, M. G., Smiley, J. R., Thomas, C., & Saif, L. J. (2004). Genetic recombination between two genotypes of genogroup III bovine noroviruses (BoNVs) and capsid sequence diversity among BoNVs and Nebraska-like bovine enteric caliciviruses. *Journal of Clinical Microbiology, 42*(11), 5214–5224; Koopmans, M., & Duizer, E. (2004). Foodborne viruses: An emerging problem. *International Journal of Food Microbiology, 90*(1), 23–41. https://doi.org/10.1128/JCM.42.11.5214-5224.2004

Hanlon, P., & Sewalt, V. (2021). GEMs: Genetically engineered microorganisms and the regulatory oversight of their uses in modern food production. *Critical Reviews in Food Science and Nutrition, 61*(6), 959–970. https://doi.org/10.1080/10408398.2020.1749026

Hassoun, A., Anusha Siddiqui, S., Smaoui, S., Ucak, İ., Arshad, R. N., Bhat, Z. F., Bhat, H. F., Carpena, M., Prieto, M. A., Aït-Kaddour, A., Pereira, J. A. M., Zacometti, C., Tata, A., Ibrahim, S. A., Ozogul, F., & Camara, J. S. (2024). Emerging technological advances in improving the safety of muscle foods: Framing in the context of the food revolution 4.0. *Food Reviews International, 40*(1), 37–78. https://doi.org/10.1080/87559129.2022.2149776

Havelaar, A. H., et al. (2010). Future challenges to microbial food safety. *International Journal of Food Microbiology, 139*, S79–S94. https://doi.org/10.1016/j.ijfoodmicro.2009.10.015

Henson, S., & Reardon, T. (2005). Private agri-food standards: Implications for food policy and the agri-food system. *Food Policy, 30*(3), 241–253. https://doi.org/10.1016/j.foodpol.2005.05.002

Hoffmann, S., Ashton, L., & Ahn, J. W. (2021). Food safety: A policy history and introduction to avenues for economic research. *Applied Economic Perspectives and Policy, 43*(2), 680–700. https://doi.org/10.1002/aepp.13158

Høiby, N. (2024). Louis Pasteur and the birth of microbiology in Denmark. *APMIS, 132*(1), 43–52. https://doi.org/10.1111/apm.13279

Ibrahim Mohamed, M.-Y. (2021). Occurrence of antimicrobial resistance in foodborne bacteria (Campylobacter and E. coli): A food safety issue and public health hazard. *Nutrition and Food Science International Journal, 11*(1). https://doi.org/10.19080/NFSIJ.2021.11.555801

Ibrahim, S. A., Ayivi, R. D., Zimmerman, T., Siddiqui, S. A., Altemimi, A. B., Fidan, H., Esatbeyoglu, T., & Bakhshayesh, R. V. (2021). Lactic acid bacteria as antimicrobial agents: Food safety and microbial food spoilage prevention. In *Foods* (Vol. 10, Issue 12). MDPI. https://doi.org/10.3390/foods10123131

Ismail, T., Layla, A., & Akhtar, S. (2022). Fermented plant protein products. In *plant protein foods* (pp. 197–222). Springer International Publishing. https://doi.org/10.1007/978-3-030-91206-2_7

Jalalian, S. H., Ramezani, M., Danesh, N. M., Alibolandi, M., Abnous, K., & Taghdisi, S. M. (2018). A novel electrochemical aptasensor for detection of aflatoxin M1 based on target-induced immobilization of gold nanoparticles on the surface of electrode. *Biosensors and Bioelectronics, 117*, 487–492. https://doi.org/10.1016/j.bios.2018.06.055

Jans, C., Mulwa Kaindi, D. W., & Meile, L. (2016). Innovations in food preservation in pastoral zones. *Rev Sci Tech, 35*(2), 597–610. https://doi.org/10.20506/rst.35.2.2527

Jarvis, K. G., Daquigan, N., White, J. R., Morin, P. M., Howard, L. M., Manetas, J. E., Ottesen, A., Ramachandran, P., & Grim, C. J. (2018). Microbiomes associated with foods from plant and animal sources. *Frontiers in Microbiology, 9*(OCT), 1–13. https://doi.org/10.3389/fmicb.2018.02540

Jay, J. M., Loessner, M. J., & Golden, D. A. (2005). History of microorganisms in food. In *Modern food microbiology* (pp. 3–9). Springer US. https://doi.org/10.1007/0-387-23413-6_1

Jaykus, L. A., et al. (2019). Next-generation technologies for foodborne pathogen detection. *Annual Review of Food Science and Technology, 10*, 425–447. https://doi.org/10.1146/annurev-food-032818-121328

Jia, W.-L et al. (2023). Antibiotics in soil and water: Occurrence, fate, and risk. *Current Opinion in Environmental Science & Health, 32*, 100437. https://doi.org/10.1016/j.coesh.2022.100437

Karanth, S. et al. (2023). Linking microbial contamination to food spoilage and food waste: the role of smart packaging, spoilage risk assessments, and date labelling. *Frontiers in Microbiology, 14*. https://doi.org/10.3389/fmicb.2023.1198124

Kasimanickam, V., Kasimanickam, M., & Kasimanickam, R. (2021). Antibiotics use in food animal production: escalation of antimicrobial resistance: Where are we now in combating AMR? *Medical Sciences, 9*(1), 14. https://doi.org/10.3390/medsci9010014

Kariyawasam, K. M. G. M., Lee, N.-K., & Paik, H.-D. (2021). Fermented dairy products as delivery vehicles of novel probiotic strains isolated from traditional fermented Asian foods. *Journal of Food Science and Technology, 58*(7), 2467–2478. https://doi.org/10.1007/s13197-020-04857-w

Khayeka-Wandabwa, C., Choge, J. K., Linnemann, A. R., & Schoustra, S. (2024). Linking fermented foods to microbial composition and valorisation: Blueprint for Kenya. *Food Reviews International*, 1–21. https://doi.org/10.1080/87559129.2024.2355992

Kishore, A., Kumar, P., Singh, A., Kumari, K., & Kumar, N. (2024). Innovative packaging strategies for freshness and safety of food products: A review. *Packaging Technology and Science, 37*(5), 399–427. https://doi.org/10.1002/pts.2800

Koirala, S., & Anal, A. K. (2021). Probiotics-based foods and beverages as future foods and their overall safety and regulatory claims. *Future Foods, 3*, Article 100013. https://doi.org/10.1016/j.fufo.2021.100013

Koopmans, M., & Duizer, E. (2004). Foodborne viruses: An emerging problem. *International Journal of Food Microbiology, 90*(1), 23–41. https://doi.org/10.1016/S0168-1605(03)00169-7

Kumar, A., Zhimo, V. Y., Biasi, A., Feygenberg, O., Salim, S., Wisniewski, M., & Droby, S. (2021). Impact of packhouse treatments on the peel microbiome of mandarin fruit (cv. Orr). *Postharvest Biology and Technology, 176*, 111519. ISSN 0925–5214. https://doi.org/10.1016/j.postharvbio.2021.111519

Kumar, V., Singh, K., Shah, M. P., Singh, A. K., Kumar, A., & Kumar, Y. (2021). Application of Omics technologies for microbial community structure and function analysis in contaminated environment. In *Wastewater treatment* (pp. 1–40). Elsevier. https://doi.org/10.1016/B978-0-12-821881-5.00001-5

Kursa, W., Jamiołkowska, A., Wyrostek, J., & Kowalski, R. (2022). Antifungal effect of plant extracts on the growth of the cereal pathogen fusarium spp.—an in vitro study. *Agronomy, 12*(12), 3204. https://doi.org/10.3390/agronomy12123204

Lee, S., Han, A., Yoon, J.-H., & Lee, S.-Y. (2022). Growth evaluation of Escherichia coli O157:H7, Salmonella typhimurium, and Listeria monocytogenes in fresh fruit and vegetable juices via predictive modeling. *LWT, 162,* 113485. https://doi.org/10.1016/j.lwt.2022.113485

Li, A. Y. et al. (2014). Potential for dissemination of Phytophthora cinnamomi by feral pigs via ingestion of infected plant material. *Biological Invasions, 16*(4), 765–774. https://doi.org/10.1007/s10530-013-0535-7

Lianou, A., Nychas, G.-J. E., & Koutsoumanis, K. P. (2020). Strain variability in biofilm formation: A food safety and quality perspective. *Food Research International, 137,* 109424. https://doi.org/10.1016/j.foodres.2020.109424

Liu, L., Chen, X., Hao, L., Zhang, G., Jin, Z., Li, C., Yang, Y., Rao, J., & Chen, B. (2022). Traditional fermented soybean products: processing, flavor formation, nutritional and biological activities. *Critical Reviews in Food Science and Nutrition, 62*(7), 1971–1989. https://doi.org/10.1080/10408398.2020.1848792

Lorenzo, J. M. et al. (2018). Main groups of microorganisms of relevance for food safety and stability: General aspects and overall description. In *Innovative technologies for food preservation: Inactivation of spoilage and pathogenic microorganisms* (pp. 53–107). Elsevier. https://doi.org/10.1016/B978-0-12-811031-7.00003-0

Lourenco, A. A., Burgess, C., & Ells, T. (2022). Foodborne pathogens and food safety. In Lourenco, A.A., Burgess, C., & Ells. T (Eds.), *MDPI.* https://doi.org/10.3390/books978-3-0365-4431-1

Madilo, F. K. et al. (2023). Foodborne pathogens awareness and food safety knowledge of street-vended food consumers: A case of university students in Ghana. *Heliyon, 9*(7). https://doi.org/10.1016/j.heliyon.2023.e17795

Mahmud, K. et al. (2020). Current progress in nitrogen fixing plants and microbiome research. *Plants. MDPI AG.* https://doi.org/10.3390/plants9010097

Mancuso, G., et al. (2021). Bacterial antibiotic resistance: the most critical pathogens. *Pathogens, 10*(10), 1310. https://doi.org/10.3390/pathogens10101310

Martella, V., Campolo, M., Lorusso, E., Cavicchio, P., Camero, M., Bellacicco, A. L., Decaro, N., Elia, G., Greco, G., Corrente, M., Desario, C., Arista, S., Banyai, K., Koopmans, M., & Buonavoglia, C. (2007). Norovirus in captive lion cub (Panthera leo). *Emerging Infectious Diseases, 13*(7), 1071–1073. https://doi.org/10.3201/eid1307.070268

Mather, A. E., Gilmour, M. W., Reid, S. W. J., & French, N. P. (2024). Foodborne bacterial pathogens: Genome-based approaches for enduring and emerging threats in a complex and changing world. *Nature Reviews Microbiology.* https://doi.org/10.1038/s41579-024-01051-z

Melara, E. G., Avellaneda, M. C., Valdivié, M., García-Hernández, Y., Aroche, R., & Martínez, Y. (2022). Probiotics: Symbiotic relationship with the animal host. *Animals, 12*(6), 719. https://doi.org/10.3390/ani12060719

Eruaga, M. A. (2024). Policy strategies for managing food safety risks associated with climate change and agriculture. *International Journal of Scholarly Research and Reviews, 4*(1), 021–032. https://doi.org/10.56781/ijsrr.2024.4.1.0026

Mirres, A. C. D. M., Silva, B. E. P. D. M. D., Tessaro, L., Galvan, D., Andrade, J. C. D., Aquino, A., Joshi, N., & Conte-Junior, C. A. (2022). Recent advances in nanomaterial-based biosensors for pesticide detection in foods. *Biosensors, 12*(8), 572. https://doi.org/10.3390/bios12080572

Montagnini, F., & Metzel, R. (2024). The contribution of agroforestry to sustainable development goal 2: End hunger, Achieve food security and improved nutrition, and promote sustainable agriculture. In *Integrating landscapes: agroforestry for biodiversity conservation and food sovereignty* (pp. 21–67). https://doi.org/10.1007/978-3-031-54270-1_2

Mousavi Khaneghah, A., Abhari, K., Eş, I., Soares, M. B., Oliveira, R. B. A., Hosseini, H., Rezaei, M., Balthazar, C. F., Silva, R., Cruz, A. G., Ranadheera, C. S., & Sant'Ana, A. S. (2020). Interactions between probiotics and pathogenic microorganisms in hosts and foods: A review. *Trends in Food Science & Technology, 95,* 205–218. https://doi.org/10.1016/j.tifs.2019.11.022

Murrell, K. D., & Pozio, E. (2011). Worldwide occurrence and impact of Human Trichinellosis, 1986–2009. *Emerging Infectious Diseases, 17*(12), 2194–2202. https://doi.org/10.3201/eid1712.110896

Naresh, V., & Lee, N. (2021). A review on biosensors and recent development of nanostructured materials-enabled biosensors. *Sensors, 21*(4), 1109. https://doi.org/10.3390/s21041109

Nath, S. (2024). Advancements in food quality monitoring: Integrating biosensors for precision detection. *Sustainable Food Technology*. https://doi.org/10.1039/D4FB00094C

Nayak, R., & Waterson, P. (2019). Global food safety as a complex adaptive system: key concepts and future prospects. *Trends in Food Science and Technology*. https://doi.org/10.1016/j.tifs.2019.07.040

Nesakumar, N., Lakshmanakumar, M., Srinivasan, S., Jayalatha Jbb, A., & Balaguru Rayappan, J. B. (2021). Principles and recent advances in biosensors for pathogens detection. *ChemistrySelect, 6*(37), 10063–10091. https://doi.org/10.1002/slct.202101062

Pantaleo, G., Correia, B., Fenwick, C., Joo, V. S., & Perez, L. (2022). Antibodies to combat viral infections: Development strategies and progress. *Nature Reviews Drug Discovery, 21*(9), 676–696. https://doi.org/10.1038/s41573-022-00495-3

Patarata, L., Munekata, P. E. S., & Fraqueza, M. J. dos R. (2024). Probiotics as starter and nonstarter cultures in fermented foods. In *Strategies to improve the quality of foods* (pp. 203–233). Elsevier. https://doi.org/10.1016/B978-0-443-15346-4.00008-2

Patel, P., Kumar, S., & Kumar, A. (2022a). Plant microbiota: a prospect to Edge off postharvest loss. In *Biocontrol mechanisms of endophytic microorganisms* (pp. 261–284). Elsevier. https://doi.org/10.1016/B978-0-323-88478-5.00006-7

Patel, P., Kumar, S., & Kumar, A. (2022b). Plant microbiota: a prospect to Edge off postharvest loss. In *Biocontrol mechanisms of endophytic microorganisms* (pp. 261–284). Elsevier. https://doi.org/10.1016/B978-0-323-88478-5.00006-7

Pawlowski, Z. S., Pozio, E., & Schultz, M. G. (2012). Epidemiology and control of Trichinellosis in the European Union. *Parasitology Research, 110*(2), 557–570. https://doi.org/10.1007/s00436-011-2585-7

Pei, X., et al. (2011). The China melamine milk scandal and its implications for food safety regulation. *Food Policy, 36*(3), 412–420. https://doi.org/10.1016/j.foodpol.2011.03.008

Paz-Arteaga, S. L., Ascacio-Valdés, J. A., Aguilar, C. N., Cadena-Chamorro, E., Serna-Cock, L., Aguilar-González, M. A., Ramírez-Guzmán, N., & Torres-León, C. (2023). Bioprocessing of pineapple waste for sustainable production of bioactive compounds using solid-state fermentation. *Innovative Food Science & Emerging Technologies, 85,*103313. https://doi.org/10.1016/j.ifset.2023.103313

Peng, M., Tabashsum, Z., Anderson, M., Truong, A., Houser, A. K., Padilla, J., Akmel, A., Bhatti, J., Rahaman, S. O., & Biswas, D. (2020). Effectiveness of probiotics, prebiotics, and prebiotic-like components in common functional foods. *Comprehensive Reviews in Food Science and Food Safety, 19*(4), 1908–1933. https://doi.org/10.1111/1541-4337.12565

Piqué, N. et al. (2015). Virulence factors of Erwinia amylovora: A review. *International Journal of Molecular Sciences. MDPI AG*, 12836–12854. https://doi.org/10.3390/ijms160612836

Primožič, M., Knez, Ž, & Leitgeb, M. (2021). (Bio) Nanotechnology in food science—food packaging. *Nanomaterials, 11*(2), 292. https://doi.org/10.3390/nano11020292

Probst, C., Njapau, H. & Cotty, P. J. (2007). Outbreak of an Acute Aflatoxicosis in Kenya in 2004: Identification of the causal agent. *Applied and Environmental Microbiology, 73*(8), 2762–2764.

Qu, Z. et al. (2024). *Fusarium mycotoxins: The major food contaminants*. mLife [Preprint]. https://doi.org/10.1002/mlf2.12112

Rabetafika, H. N., Razafindralambo, A., Ebenso, B., & Razafindralambo, H. L. (2023). Probiotics as antibiotic alternatives for human and animal applications. *Encyclopedia, 3*(2), 561–581. https://doi.org/10.3390/encyclopedia3020040

Ray, M. R., et al. (2005). Respiratory and general health impairments of workers employed in a municipal solid waste disposal at an open landfill site in Delhi. *International Journal of Hygiene and Environmental Health, 208*(4), 255–262. https://doi.org/10.1016/j.ijheh.2005.02.001

Rhouma, M., Gaucher, M.-L., Badredine, S., Bekal, S., & Sanders, P. (2024). Food risk assessment in the farm-to-table continuum: Report from the conference on good hygiene practices to ensure food safety. *Agriculture & Food Security, 13*(1), 21. https://doi.org/10.1186/s40066-024-00476-2

Ripolles-Avila, C., Martínez-Garcia, M., Capellas, M., Yuste, J., Fung, D. Y. C., & Rodríguez-Jerez, J. (2020). From hazard analysis to risk control using rapid methods in microbiology: A practical approach for the food industry. *Comprehensive Reviews in Food Science and Food Safety, 19*(4), 1877–1907. https://doi.org/10.1111/1541-4337.12592

Rosario, A. I. L. S., Castro, V. S., Santos, L. F., Lisboa, R. C., Vallim, D. C., Silva, M. C. A., Figueiredo, E. E. S., Conte-Junior, C. A., & Costa, M. P. (2021). Shiga toxin–producing *Escherichia coli* isolated from pasteurized dairy products from Bahia, Brazil. *Journal of Dairy Science, 104*(6), 6535–6547.https://doi.org/10.3168/jds.2020-19511

Rutter, W. B., Franco, J., & Gleason, C. (2022). Rooting out the mechanisms of root-knot nematode-plant interactions. https://doi.org/10.1146/annurev-phyto-021621

Santi, R., et al. (2022). Sustainable food packaging: An integrative framework. *Sustainability, 14*(13), 8045. https://doi.org/10.3390/su14138045

Scallan, E., et al. (2011). Foodborne illness acquired in the United States—Major pathogens. *Emerging Infectious Diseases.* https://doi.org/10.3201/eid1701.P11101

Scharff, R. L. (2012). Economic burden from health losses due to foodborne illness in the United States. *Journal of Food Protection, 75*(1), 123–131. https://doi.org/10.4315/0362-028X.JFP-11-058

Sehnal, L. et al. (2021). *Microbiome composition and function in aquatic vertebrates: small organisms making big impacts on aquatic animal health.* Frontiers in Microbiology. Frontiers Media S.A. https://doi.org/10.3389/fmicb.2021.567408

Sequino, G., Valentino, V., Villani, F., & De Filippis, F. (2022). Omics-based monitoring of microbial dynamics across the food chain for the improvement of food safety and quality. *Food Research International, 157*, Article 111242. https://doi.org/10.1016/j.foodres.2022.111242

Sergaki, C. et al. (2018). Challenges and approaches in Microbiome research: From fundamental to applied. *Frontiers in Plant Science, 9*. https://doi.org/10.3389/fpls.2018.01205

Shi, T., Chen, H., Liu, Y., Wu, Y., & Lin, F. (2024). Clinical applications of metagenomic next-generation sequencing in the identification of pathogens in periprosthetic joint infections: A retrospective study. *Journal of Orthopaedic Surgery and Research, 19*(1), 301. https://doi.org/10.1186/s13018-024-04745-5

Shwaiki, L. N., Lynch, K. M., & Arendt, E. K. (2021). Future of antimicrobial peptides derived from plants in food application–a focus on synthetic peptides. *Trends in Food Science & Technology, 112*, 312–324. https://doi.org/10.1016/j.tifs.2021.04.010

Silva, A., et al. (2024). Listeria monocytogenes from Food Products and Food Associated Environments: Antimicrobial resistance, genetic clustering and biofilm insights. *Antibiotics, 13*(5), 447. https://doi.org/10.3390/antibiotics13050447

Singh, J., Roychoudhury, A., Srivastava, M., Solanki, P. R., Lee, D. W., Lee, S. H., & Malhotra, B. D. (2013). A highly efficient rare earth metal oxide nanorods based platform for aflatoxin detection. *Journal of Materials Chemistry B, 1*(35), 4493–4503. https://doi.org/10.1039/C3TB20690D

Singh, R. P., & Desrosier. (2024). *Food preservation.* Encyclopedia Britannica. https://www.britannica.com/topic/food-preservation (Accessed: 16 June 2024).

Smyth, S. J., Webb, S. R., & Phillips, P. W. B. (2021). The role of public-private partnerships in improving global food security. *Global Food Security, 31*, Article 100588. https://doi.org/10.1016/j.gfs.2021.100588

Snyder, A. B., Martin, N., & Wiedmann, M. (2024). Microbial food spoilage: Impact, causative agents and control strategies. *Nature Reviews Microbiology.* https://doi.org/10.1038/s41579-024-01037-x

Stewart, G. F., & Amerine, M. A. (2012). *Introduction to food science and technology.*

Suganya, R. et al. (2019). Mucormycosis: A brief review. *Journal of Pure and Applied Microbiology*, 161–165. https://doi.org/10.22207/JPAM.13.1.16

Sun, W., Shahrajabian, M. H., & Lin, M. (2022). Research progress of fermented functional foods and protein factory-microbial fermentation technology. *Fermentation, 8*(12), 688. https://doi.org/10.3390/fermentation8120688

Surya, R., & Nugroho, D. (2023). Kimchi throughout millennia: a narrative review on the early and modern history of kimchi.*Journal of Ethnic Foods, 10*(1), 5. https://doi.org/10.1186/s42779-023-00171-w

Tabari, H. (2020). Climate change impact on flood and extreme precipitation increases with water availability. *Scientific Reports*, 10(1), 13768. https://doi.org/10.1038/s41598-020-70816-2

Taiwo, O. R., Onyeaka, H., Oladipo, E. K., Oloke, J. K., & Chukwugozie, D. C. (2024). Advancements in predictive microbiology: integrating new technologies for efficient food safety models. *International Journal of Microbiology, 2024*, 1–21. https://doi.org/10.1155/2024/6612162

Tanner, F. W. (1994). *The microbiology of foods* (2nd edn)

Tarasova, I. A., Finogenova, T. A., & Gurieva, K. B. (2022). Study of the process of tea moisture sorption. *Tovaroved prodovolstvennykh tovarov (Commodity specialist of food products)*, (8), 514–521. https://doi.org/10.33920/igt-01-2208-03

Tay, C. S., Yeo, Y. T., Ng, K. R., Yap, P. K., & Chen, W. N. (2023). Metabolomics-driven comparison of the nutritional and functional food characteristics of postbiotic and probiotic okara. *ACS Food Science & Technology, 3*(7), 1165–1174. https://doi.org/10.1021/acsfoodscitech.3c00050

Tenhagen, B.-A., Flor, M., Alt, K., Knüver, M.-T., Buhler, C., Käsbohrer, A., & Stingl, K. (2021). Association of antimicrobial resistance in campylobacter spp. in broilers and Turkeys with antimicrobial use. *Antibiotics, 10*(6), 673. https://doi.org/10.3390/antibiotics10060673

Tompkin, R. B. (2001). Interactions between government and industry food safety activities. *Food Control, 12*(4), 203–207. https://doi.org/10.1016/S0956-7135(00)00038-4

Tripathi, J. (2021). Food safety and methods to ensure food security in the face of climate change. *CAB Reviews: Perspectives in Agriculture, Veterinary Science, Nutrition and Natural Resources, 16*(015). https://doi.org/10.1079/PAVSNNRYR202116015

Trollman, H., Bader, F., Luo, Y., Garcia-Garcia, G., Jagtap, S., Parra-López, C., Colwill, J., & Trollman, F. (2024). Food safety 4.0. In *Food Industry 4.0* (pp. 61–81). Elsevier. https://doi.org/10.1016/B978-0-443-15516-1.00004-9

Upadhyay, A. K. (2020). A microbial menace: Emerging infections in the 21st century. *Cohesive Journal of Microbiology & Infectious Disease, 4*(2). https://doi.org/10.31031/CJMI.2020.03.000583

Van der Meulen, B. (2010). The global arena of food law: emerging contours of a meta-framework. *Erasmus Law Review, 3*(4), 217–240. https://doi.org/10.5553/ELR221026712010003004003

Vasile, C., & Baican, M. (2021). Progresses in food packaging, food quality, and safety—Controlled-release antioxidant and/or antimicrobial packaging. *Molecules, 26*(5), 1263. https://doi.org/10.3390/molecules26051263

Wang, H., Huang, J., & Jiang, X. (2023). Perspectives on using a competitive exclusion approach to control listeria monocytogenes in Biological Soil Amendments of Animal Origin (BSAAO): A review. *Applied Microbiology, 3*(3), 786–804. https://doi.org/10.3390/applmicrobiol3030055

Werner, S., Maschke, R. W., Eibl, D., & Eibl, R. (2018). Bioreactor technology for sustainable production of plant cell-derived products. In *Bioprocessing of plant in vitro systems* (pp. 413–432). https://doi.org/10.1007/978-3-319-54600-1_6

Williams, M. S., Ebel, E. D., Saini, G., & Nyirabahizi, E. (2020). Changes in salmonella contamination in meat and poultry since the introduction of the pathogen reduction and hazard analysis and critical control point rule. *Journal of Food Protection, 83*(10), 1707–1717. https://doi.org/10.4315/JFP-20-126

World Health Organization. (2022). *Food safety*. https://www.who.int/news-room/fact-sheets/detail/food-safety (Accessed: 4 June 2024).

Wu, D., Elliott, C., & Wu, Y. (2021). Food safety strategies: the one health approach to global challenges and China's actions. *China CDC Weekly, 3*(24), 507–513. https://doi.org/10.46234/ccdcw2021.131

Xu, J., Peng, S., Xiong, Y., Zheng, Z., Liu, M., Xu, J., Chen, W., Liu, M., Kong, J., Wang, C., Wang, Y., Rao, L., Zhao, L., & Liao, X. (2024). A review on fermented vegetables: Microbial community and potential upgrading strategy via inoculated fermentation. *Comprehensive Reviews in Food Science and Food Safety, 23*(3). https://doi.org/10.1111/1541-4337.13362

Xue, Y., Gao, Y., Guo, M., Zhang, Y., Zhao, G., Xia, L., Ma, J., Cheng, Y., Wang, H., Sun, J., Wang, Z., & Yan, Y. (2024). Phage cocktail superimposed disinfection: A ecological strategy for preventing pathogenic bacterial infections in dairy farms. *Environmental Research, 252*, Article 118720. https://doi.org/10.1016/j.envres.2024.118720

Yan, C., Wang, Q., Yang, Q., & Wu, W. (2020). Recent advances in aflatoxins detection based on nanomaterials. *Nanomaterials, 10*(9), 1626. https://doi.org/10.3390/nano10091626

Yang, Z., Chambers, H., DiCaprio, E., Gao, G., & Li, J. (2018). Internalization and dissemination of human norovirus and Tulane virus in fresh produce is plant dependent. *Food Microbiology, 69*, 25–32. https://doi.org/10.1016/j.fm.2017.07.015

Yogi, L. N. et al. (2024). An overview on the impact of genetically engineered organisms on crop yield and safety. *Archives of Agriculture and Environmental Science, 9*(1), 175–179. https://doi.org/10.26832/24566632.2024.0901025

Yoo, S., Jung, S.-C., Kwak, K., & Kim, J.-S. (2024). The role of prebiotics in modulating gut microbiota: Implications for human health. *International Journal of Molecular Sciences, 25*(9), 4834. https://doi.org/10.3390/ijms25094834

Zahra, S. T. et al. (2023) 'Probing the potential of salinity-tolerant endophytic bacteria to improve the growth of mungbean [Vigna radiata (L.) Wilczek]'. *Frontiers in Microbiology, 14*. https://doi.org/10.3389/fmicb.2023.1149004

Zhao, X. et al. (2017). Biofilm formation and control strategies of foodborne pathogens: food safety perspectives. *RSC Advances, 7*(58), 36670–36683. https://doi.org/10.1039/C7RA02497E

Part III
Sustainable Solutions and Innovations

Chapter 7
Sustainable Agriculture: Role and Importance

T. S. Archana, Devendra Kumar, Velugolu Venkata Suneel Kumar, and Gagan Kumar

Abstract Sustainable agriculture is increasingly recognized as a crucial solution to the pressing challenges of global food security, environmental degradation, and climate change. This chapter delves into the significance and role of sustainable agricultural practices in fostering resilient, productive, and environmentally sound farming systems. It emphasizes the need to balance increasing food production to meet the growing population's demands and preserving essential ecological functions. Central to sustainable agriculture are practices that enhance soil fertility, conserve water resources, and promote biodiversity while minimizing the use of synthetic inputs like chemical fertilizers and pesticides. Methods such as crop rotation, organic farming, conservation tillage, agroforestry, and integrated pest management (IPM) are highlighted for their contributions to improving soil structure, reducing erosion, enhancing nutrient cycling, and controlling pests through natural mechanisms. These practices reduce dependence on external inputs, lower greenhouse gas emissions, and support long-term agricultural productivity. Sustainable farming systems often incorporate traditional knowledge, aligning with local conditions and resources, increasing productivity and strengthening community resilience. In addition, the shift toward sustainable practices can reduce poverty by enhancing access to markets, improving crop quality, and increasing farmer income opportunities. Moreover, sustainable agriculture fosters a more equitable distribution of resources, ensuring that future generations have access to fertile land, clean water, and a healthy environment. It advocates integrating innovative technologies, such as precision agriculture and biotechnologies, with traditional methods to enhance efficiency and adaptability in farming systems.

Keywords Sustainability · Integrated pest management · Ecosystem · Soil fertility · Nutrient

T. S. Archana · D. Kumar · V. V. S. Kumar
Department of Plant Pathology, School of Agriculture, Lovely Professional University (LPU), Phagwara, Punjab, India

G. Kumar (✉)
Department of Plant Pathology, Acharya Narendra Deva University of Agriculture & Technology, Kumarganj, Ayodhya, Uttar Pradesh, India
e-mail: gkcpp@nduat.org

J. Parray and A. K. Haghi (eds.), *Food and Water Security*,
https://doi.org/10.1007/978-3-032-04870-7_7

7.1 Introduction

The need for more food production to feed the rapidly growing global population has posed a significant risk to agricultural sustainability. Climate change also significantly challenges global food security by negatively affecting agricultural output (yield), particularly in developing countries. High-input farming (*intensified agricultural)* to increase food production from available cropland has put the ecological balance at risk due to higher emissions levels (Rehman et al. (2022). It also led to erosion of biodiversity, soil health, and environmental sustainability due to excess use of chemical fertilizers and pesticides. So, there is a need to adopt a multidimensional approach to agriculture sustainability without damaging social, economic, and environmental integrity.

Sustainable agriculture is one of the holistic approaches in farming that balance economic viability, environmental stability and social equity (Bathaei & Streimikiene (2023)). It seeks to meet current agricultural needs without compromising the ability of future generations to meet their own needs. Sustainable agriculture balances productivity by conserving resources, minimizing adverse environmental effects, and promoting resilient and inclusive agricultural practices. Sustainable farming practices have deep historical roots that evolved through centuries of agricultural innovation, responding to various food production challenges, resource conservation, and environmental stewardship. The development of sustainable agriculture can be traced through distinct periods that shaped modern approaches to agricultural sustainability.

7.1.1 Early Agricultural Practices and Indigenous Knowledge

The earliest forms of agriculture were subsistence-based, relying on indigenous knowledge and a deep understanding of local ecosystems. Early farmers used crop rotation, polyculture, and agroforestry for thousands of years to maintain soil fertility and ensure food security. Indigenous cultures worldwide practiced slash-and-burn techniques managed to graze, and developed intricate irrigation systems, adapting farming methods to local environmental conditions (Smith & Anderson, 1994). These early methods were inherently sustainable, relying on biodiversity and ecological balance to minimize environmental impact and optimize resource use.

7.1.2 The Industrial Revolution and Agricultural Intensification

With the onset of the Industrial Revolution in the eighteenth and nineteenth centuries, agriculture became more mechanized and intensified. Introducing synthetic fertilizers, chemical pesticides, and large-scale monocropping practices marked a shift towards increased productivity, but at the cost of environmental health. This period, often called the Green Revolution (thc 1940s–the 1960s), significantly boosted food production, particularly in developing countries, but also led to widespread soil degradation, water scarcity, and loss of biodiversity (Pretty, 2008). The reliance on non-renewable resources, such as fossil fuels for machinery and fertilizers, contributed to the degradation of natural ecosystems, creating a need for more sustainable approaches.

7.1.3 Emergence of Modern Sustainable Agriculture

In response to the environmental degradation caused by industrial agriculture, sustainable agriculture began to take shape in the twentieth century. The 1970s witnessed a growing awareness of the ecological impacts of modern farming, leading to movements advocating for organic agriculture, conservation tillage, and integrated pest management (IPM). Organic farming emerged as a strong response, emphasizing the use of natural inputs and biological processes for maintaining soil health and pest control (IFOAM, 2005). The 1980s and 1990s saw the formalization of sustainable agriculture as an academic discipline and policy goal. Key frameworks such as agroecology and permaculture were developed, promoting holistic approaches that integrate farming's social, economic, and environmental dimensions. For instance, agroecology advocates for applying ecological principles to agricultural systems, emphasizing biodiversity, ecosystem services, and the empowerment of small-scale farmers (Altieri, 1995). Today, sustainable agriculture continues to evolve, driven by the challenges of climate change, food security, and the depletion of natural resources. Technological advancements, such as precision agriculture and using renewable energy in farming, are being integrated into sustainable practices. The focus has expanded to include social and economic sustainability, emphasizing fair labor practices, food justice, and resilience against global food system shocks (Kremen & Miles, 2012).

7.2 Sustainable Agriculture and Food Security

According to India's recent official census data and population projections, the population in 2050 will be expected to reach approximately 1.52–1.6 billion people. This population growth rate will need to increase food supply by 60–70% more than present. Moreover, the change in climate over the decades has led to heat waves due to constant increases in temperature, atmospheric CO_2 level, and repeated occurrence of dry spells in parts of the country while other places face high precipitation. These variations will significantly impact agricultural sustainability. In addition, anthropogenic activities significantly impact agriculture's sustainability due to the excess use of resources. Excess use of the resources will lead to gradual depletion of the resources. Furthermore, this will lead to lower agricultural yields and a threat to food security and safety. Food security is physical, social, and economic accessibility of healthy, safe, and nutritious food to the people at all times, which can fulfil their dietary requirements and food preference for healthy and active life (FAO, 2001). However, agricultural sustainability is crucial to accomplishing food security, as the primary object of food security is the availability of food supply. Food availability is related to crop production, which depends on the agro-climatic condition of the crop. To meet the demands of the increasing population for food security, the people focused on agriculture extension and intensification. (Adelaja & George, 2021)) Agricultural intensification increases agricultural productivity per unit of land by using more inputs such as fertilizers, irrigation, improved seeds, pesticides, and mechanization. It involves enhancing farming practices to produce more food without expanding the land used for cultivation. This intensive farming will increase food production and productivity to achieve food security. This intensive farming without sustainability will significantly affect environmental costs, affecting yield reduction in later stages because of land degradation (Fig. 7.1).

7.3 Principles of Sustainable Agriculture

The principles of sustainable agriculture emphasize an integrative approach, where ecological, economic, and social goals converge to ensure the long-term viability of farming systems. These principles guide the development and implementation of farming practices that are environmentally sound, economically profitable, and socially equitable.

7.3.1 Ecological Balance and Biodiversity Preservation

Sustainable agriculture prioritizes ecological balance by maintaining and enhancing above and below-ground biodiversity. Preserving diverse species within agricultural

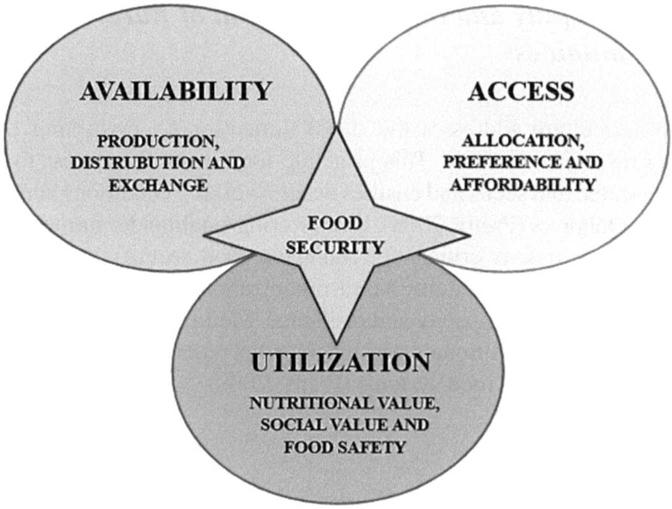

Fig. 7.1 The three components of food systems with their main elements are shown

systems, including plants, animals, and microorganisms, supports critical ecosystem services such as pollination, pest control, and nutrient cycling (Altieri, 1999). Practices like crop rotation, agroforestry, and cover cropping help maintain soil health and reduce the need for chemical inputs, thus promoting a resilient ecosystem. By fostering biodiversity, sustainable farming mitigates the risk of pest outbreaks and disease while enhancing the adaptability of agricultural systems to changing environmental conditions (Perfecto & Vandermeer, 2010).

7.3.2 Economic Viability and Resilience of Farming Systems

For agriculture to be sustainable, it must also be economically viable, allowing farmers to earn a livelihood while maintaining the integrity of their land and resources. This involves adopting practices that increase productivity in the long run without overexploiting natural resources. Sustainable farming seeks to reduce input costs by minimizing dependence on synthetic fertilizers, pesticides, and non-renewable energy sources. Instead, it emphasizes organic inputs, efficient water management, and the use of renewable energy (Gliessman, 2014). Economic resilience is also enhanced through the diversification of crops and income sources, reducing farmers' vulnerability to market fluctuations and environmental changes.

7.3.3 Social Equity and the Empowerment of Rural Communities

Sustainable agriculture addresses the social dimension by promoting equity and empowering rural communities. This principle focuses on fair access to resources such as land, water, and seeds and ensures decent working conditions and fair wages for agricultural laborers (Pretty, 2008). Empowering smallholder farmers, especially in developing countries, is critical for achieving food security and sustainability. Programs that support community-based organizations, cooperatives, and farmer education help build local capacity and resilience. Moreover, sustainable agriculture often seeks to preserve traditional knowledge and practices, recognizing their value in maintaining sustainable food systems (Pretty, 2008).

7.3.4 Minimizing Resource Depletion (Soil, Water, Energy)

Sustainable farming is inherently focused on efficiently using natural resources to minimize depletion. Soil health is a key concern, with practices like conservation tillage, organic fertilization, and biochar helping to enhance soil structure, increase organic matter, and improve water retention (Lal, 2004). Water conservation is also prioritized through drip irrigation, rainwater harvesting, and drought-resistant crop varieties. In terms of energy, sustainable agriculture advocates for using renewable energy sources like solar and wind while reducing the carbon footprint of farming operations (Tilman et al., 2002).

7.4 Sustainable Agriculture and Food Security: The Impact of Food Production

Sustainable agriculture is vital in solving global food security, especially given growing human populations, diminishing environmental conditions, and changing climate patterns. Producing sufficient nutritious food to satisfy the needs of an expanding population while reducing adverse environmental effects is necessary to ensure food security. The key objectives of sustainable practices are to increase crop yields, decrease food loss and waste, and conserve natural resources for future generations: Sustainable Agriculture and Water Security.

7.4.1 Enhancing Crop Yields Through Sustainable Practices

A key barrier to achieving food security is increasing crop production in an environmentally friendly way. The main objective of sustainable agriculture is to increase agricultural yields without diminishing the health of the soil, consuming excessive amounts of water, or producing more greenhouse gas emissions. (Liu et al. (2024)) Many approaches enhancing crop yields through sustainable practices like:

Precision agriculture leverages technologies such as drones, sensors, and GPS to monitor crop health, soil moisture, and nutrient levels, allowing farmers to use resources more efficiently. This approach reduces the need for excessive water, fertilizers, and pesticides while maximizing agricultural yield (Nowak (2021)). Additionally, crop diversification and rotation help maintain soil fertility, reduce the occurrence of pests and diseases, and promote healthier plants without the heavy use of chemicals. Improved seed varieties, bred for disease resistance, drought tolerance, and higher yields, further enhance agricultural productivity, particularly in regions facing extreme weather or water shortages. Together, these practices contribute to more sustainable farming by improving resource management and increasing resilience to environmental challenges (Lampridi et al. (2019)).

7.4.2 Minimizing Food Wastage and Loss

A large amount of food is lost or wasted before it reaches customers globally. In addition to impacting food security, this loss results in the waste of labor, land, and water—resources utilized in production. Improving food availability and maintaining agricultural systems' sustainability depends on addressing food loss and waste.

7.4.3 Causes of Food Loss and Waste

Pre-harvest losses are often driven by unpredictable weather patterns, pest infestations, and inefficient farming practices, which can significantly reduce crop yields before harvesting. Post-harvest losses occur when crops deteriorate due to inadequate transportation, storage, and processing infrastructure, leading to significant food waste. Additionally, consumer-level waste is a major issue, as food is often discarded at the retail and household levels due to over-purchasing, improper storage, and spoilage (Marimuthu et al. (2024)). These factors contribute to the loss of valuable resources throughout the food supply chain.

7.5 Impact of Food Production on the Environment

Food production dramatically impacts the environment, from greenhouse gas emissions and water consumption to land use and deforestation. Long-term food security is threatened by unsustainable farming methods, which can also result in biodiversity loss, water scarcity, and soil degradation.

7.5.1 Environmental Challenges in Food Production

Agriculture accounts for roughly 70% of global freshwater consumption, and unsustainable irrigation practices can lead to groundwater depletion and increased water scarcity. Additionally, soil degradation is a growing concern due to the overuse of synthetic fertilizers, pesticides, intensive farming, and monocropping, all of which diminish soil quality and its ability to retain moisture and sustain healthy crops. Agriculture also significantly contributes to greenhouse gas emissions, mainly methane and nitrous oxide (Salimi et al. (2021)), primarily released through livestock farming. These emissions significantly accelerate global warming, highlighting the environmental challenges associated with conventional agricultural practices.

7.5.2 Sustainable Agriculture and Water Security

Freshwater is an essential natural resource often wasted in many ways, leading to adverse ecological consequences. In developing countries, agriculture consumes 90% of available water to produce 40% of the world's food. Enhancing crop productivity through effective irrigation methods like micro irrigation techniques (drip irrigation, sprinkle irrigation, etc.) will help save much water by avoiding water losses through infiltration or runoff. Over the past 5 decades, water consumption has increased by four times, leading to straining water resources (Kumar et al. (2024)). Countries have introduced numerous measures to protect freshwater and promote its sustainable use. However, irrigation can negatively affect semi-arid regions where groundwater extraction can lead to unsustainable water use. Global water distribution is uneven, with areas experiencing excessive rainfall or drought (Michel (2023)). The FAO has initiated "*More crop per drop*" to increase the water use efficiency in agriculture to meet the food demand for the growing population without additional water resources.

7.5.3 Water Use Efficiency

Water Use Efficiency (WUE) refers to the crop yield or biomass produced per water unit. It measures how efficiently water is converted into plant growth (Callejas Moncaleano et al. (2021)). The higher the WUE, the more efficiently plants use water to produce biomass or yield.

There are two common forms of WUE:

1. **Agronomic Water Use Efficiency (AWUE)**: The crop yield ratio to the total amount of water used (irrigation plus rainfall).

$$\text{AWUE} = \text{CropYield} / \text{Water Applied}$$

2. **Physiological Water Use Efficiency (PWUE)**: The ratio of carbon fixed (photosynthesis) to the amount of water lost through transpiration.

Drip irrigation involves delivering water directly to the root zone of plants through a system of pipes and emitters, ensuring efficient water use. Subsurface irrigation takes a similar approach but delivers water below the soil surface through buried pipes, targeting the roots more precisely. In sprinkler irrigation, water is sprayed into the air and falls over the crops like rain, providing even coverage. Centre pivot irrigation, often used on large, flat fields, has a rotating sprinkler system that waters crops in a circular pattern. Furrow irrigation, on the other hand, directs water between crop rows through small trenches or furrows. Finally, surface or flood irrigation allows water to flow freely over the field's surface, saturating the soil and crops as it moves. Each method has advantages, depending on the crop type, field conditions, and water availability. Mulching involves covering the soil surface with organic or synthetic materials to minimize evaporation, suppress weed growth, and retain moisture. Rainwater harvesting is collecting and storing rainwater for agricultural use, especially valuable during water scarcity. Contour farming and terracing, often used in hilly areas, reduce water runoff and enhance water infiltration into the soil. Crop rotation and cover cropping improve soil structure, reduce erosion, and boost water retention by promoting better infiltration. Soil moisture monitoring, through sensors, allows farmers to optimize irrigation by applying water only when needed. Zero or reduced tillage practices limit soil disturbance, improving soil structure and moisture retention while reducing runoff and erosion. Additionally, growing drought-tolerant crop varieties enhances resilience to water shortages, improving water use efficiency in farming systems.

7.5.4 Challenges in Adopting Sustainable Agriculture

Adopting sustainable agriculture comes with several challenges that can hinder its widespread implementation. These challenges range from economic and policy-related issues to social and informational barriers. Understanding these obstacles is key to developing strategies to facilitate the transition towards more sustainable farming practices. One of the primary challenges in adopting sustainable agriculture is the economic cost associated with transitioning from conventional to sustainable practices. Smallholder farmers, in particular, often lack access to the necessary financial resources to invest in sustainable technologies, such as organic fertilizers, drip irrigation systems, or renewable energy sources (Pretty, 2008).

Furthermore, while beneficial in the long run, sustainable farming practices may require higher initial investments or lead to lower short-term yields, which can be financially unsustainable for farmers already operating on narrow profit margins (Scherr & McNeely, 2008). Access to credit, subsidies, and markets that reward sustainable production methods is often limited, exacerbating this challenge. In many regions, agricultural policies are designed to support conventional farming practices that prioritize productivity over sustainability. Government subsidies usually favor chemical inputs and monoculture farming, making it difficult for farmers to adopt more sustainable alternatives (Altieri, 1995). Additionally, there is often insufficient institutional support regarding technical assistance, infrastructure development, and market access for sustainably produced goods. This lack of policy alignment and institutional support poses significant obstacles for farmers willing to transition to sustainable agriculture (Gliessman, 2014). Policies that promote sustainable practices, such as payments for ecosystem services or support for organic certification, are often underdeveloped or underfunded. A significant challenge in adopting sustainable agriculture is the limited awareness and knowledge transfer among farmers, particularly in developing regions. Many farmers are unaware of the long-term benefits of sustainable practices or lack the training and education necessary to implement these practices effectively (Leeuwis & van den Ban, 2004). Extension services, crucial for disseminating knowledge and providing technical support, are often underfunded or inadequately equipped to promote sustainable farming. As a result, farmers may rely on traditional practices or choose conventional methods because they are more straightforward or more productive in the short term.

7.6 Future of Sustainable Agriculture

The future of sustainable agriculture hinges on integrating cutting-edge innovations and research with practical, scalable solutions that can address environmental and food security challenges. Advancements in technology and increased research and development are expected to play a central role in driving sustainability in agriculture and scaling it to meet the growing global food demand.

Emerging technologies, such as nanosensors, artificial intelligence (AI), and drones, are revolutionizing sustainable agriculture by enhancing precision, efficiency, and environmental stewardship. Nanosensors, for example, can monitor soil health, water quality, and the presence of pests or diseases at a microscopic level, providing real-time data that allows for precise interventions (Monreal et al., 2016). These sensors help optimize the use of inputs like water, fertilizers, and pesticides, reducing waste and minimizing the ecological footprint of farming practices. AI and machine learning algorithms are increasingly integrated into agricultural systems to predict crop yields, optimize planting schedules, and manage resources efficiently. AI-powered tools can analyze vast amounts of data collected from farms and offer actionable insights that improve real-time decision-making (Kamilaris & Prenafeta-Boldú, 2018). Similarly, drones equipped with sensors are being used to monitor crop growth, detect pest infestations, and assess water needs across large farmland areas, reducing labor costs and improving the accuracy of interventions.

Research and development (R&D) are critical to advancing sustainable agricultural practices. R&D efforts focus on improving crop resilience, developing eco-friendly inputs, and enhancing farming techniques that align with sustainability goals. Genetic research, for example, contributes to developing drought-resistant and pest-resistant crop varieties, which can thrive in challenging environmental conditions without excessive resource use (Fischer et al., 2014). Additionally, research on agroecological practices, such as regenerative agriculture and permaculture, is helping to refine approaches that build soil health, enhance biodiversity, and sequester carbon. Public and private sector investments in agricultural research are vital for continuing to develop innovative solutions that address the specific needs of smallholder farmers, particularly in developing regions. Collaborative research programs involving farmers, scientists, and policymakers can facilitate the transfer of knowledge and the adaptation of cutting-edge solutions to local conditions, ensuring that sustainable farming practices are scientifically sound and practically viable (Tittonell, 2014).

As the global population is projected to reach nearly 10 billion by 2050, scaling up sustainable agriculture is essential to ensuring food security while protecting the environment. This will require innovative technologies and the widespread dissemination of sustainable practices across diverse farming systems. Sustainable agriculture must be implemented at both small and large scales to meet this challenge, tailored to the specific needs of different regions and farming communities. Scaling up sustainable agriculture will also require supportive policies that incentivize farmers to adopt eco-friendly practices alongside efforts to create value chains that reward sustainability. Markets for sustainably produced goods must be strengthened, and consumers need to be educated on the benefits of supporting sustainable agriculture. Furthermore, international cooperation will be necessary to share technology, expertise, and resources, notably to help farmers in low-income regions where food insecurity is most pressing (Godfray et al., 2010).

7.7 Conclusion

Sustainable agriculture is crucial for ensuring the future viability of global food systems, balancing the need for food security with the imperative to protect the environment. As the global population grows, sustainable practices will be essential in mitigating resource depletion, preserving biodiversity, and addressing climate change, all while producing enough food to meet demand. This can only be achieved through a collective effort involving various stakeholders. Farmers play a pivotal role by adopting sustainable techniques such as regenerative agriculture, organic farming, and precision technologies that reduce resource use and enhance productivity. Governments must provide policy support, create financial incentives, and invest in research and infrastructure that promote sustainability. Consumers, too, have a significant impact by supporting sustainable food products and encouraging market demand for environmentally friendly farming practices. A vision for a sustainable, resilient, and equitable agricultural future involves integrating innovations in technology, such as AI and nanotechnology, with traditional knowledge and practices to create farming systems that are adaptable and resource-efficient. This future also emphasizes equity, ensuring that smallholder farmers, particularly in developing countries, have access to resources, education, and markets that enable them to thrive. By fostering collaboration among farmers, governments, and consumers, agriculture can evolve into a system that feeds the world and supports environmental health and social justice.

References

Adelaja, A., & George, J. (2021). Food and agricultural security: An introduction to the special issue. *Sustainability, 13*(21), 12129.

Altieri, M. A. (1995). *Agroecology: The science of sustainable agriculture.* Westview Press.

Altieri, M. A. (1999). The ecological role of biodiversity in agroecosystems. *Agriculture, Ecosystems & Environment, 74*(1–3), 19–31. https://doi.org/10.1016/S0167-8809(99)00028-6

Bathaei, A., & Štreimikienė, D. (2023). A systematic review of agricultural sustainability indicators. *Agriculture, 13*(2), 241.

Beddington, J. (2010). Food security: Contributions from science to a new and greener revolution. *Philosophical Transactions of the Royal Society B: Biological Sciences, 365*(1537), 61–71.

Callejas Moncaleano, D. C., Pande, S., & Rietveld, L. (2021). Water use efficiency: A review of contextual and behavioural factors. *Frontiers in Water, 3*, Article 685650.

Fischer, R. A., Byerlee, D., & Edmeades, G. O. (2014). *Crop yields and global food security: Will yield increase continue to feed the world?* Australian Centre for International Agricultural Research.

Gliessman, S. R. (2014). *Agroecology: The ecology of sustainable food systems.* CRC Press.

Godfray, H. C. J., Beddington, J. R., Crute, I. R., Haddad, L., Lawrence, D., Muir, J. F., & Toulmin, C. et al. (2010). Food security: The challenge of feeding 9 billion people. *Science, 327*(5967), 812–818. https://doi.org/10.1126/science.1185383

IFOAM. (2005). *The principles of organic agriculture.* International Federation of Organic Agriculture Movements.

Kamilaris, A., & Prenafeta-Boldú, F. X. (2018). Deep learning in agriculture: A survey. *Computers and Electronics in Agriculture, 147*, 70–90. https://doi.org/10.1016/j.compag.2018.02.016

Kremen, C., & Miles, A. (2012). Ecosystem services in biologically diversified versus conventional farming systems: Benefits, externalities, and trade-offs. *Ecology and Society, 17*(4), 40. https://doi.org/10.5751/ES-05035-170440

Kumar, K., Mukhopadhyay, K., & Thomassin, P. J. (2024). Water conservation pathways in agriculture in India to 2030: An input-output framework. *Sustainable Water Resources Management, 10*(3), 1–15.

Lal, R. (2004). Soil carbon sequestration impacts on global climate change and food security. *Science, 304*(5677), 1623–1627. https://doi.org/10.1126/science.1097396

Lampridi, M. G., Sørensen, C. G., & Bochtis, D. (2019). Agricultural sustainability: A review of concepts and methods. *Sustainability, 11*(18), 5120.

Leeuwis, C., & van den Ban, A. W. (2004). *Communication for rural innovation: Rethinking agricultural extension.* Blackwell Publishing.

Liu, M., Chen, X., & Jiao, Y. (2024). Sustainable agriculture: theories, methods. *Practices and Policies. Agriculture, 14*(3), 473.

Marimuthu, S., Saikumar, A., & Badwaik, L. S. (2024). Food losses and wastage within the food supply chain: a critical review of its generation, impact, and conversion techniques. *Waste Disposal & Sustainable Energy*, 1–16.

Meijer, S. S., Catacutan, D., Ajayi, O. C., Sileshi, G. W., & Nieuwenhuis, M. (2015). The role of knowledge, attitudes and perceptions in the adoption of agricultural and agroforestry innovations among smallholder farmers in sub-Saharan Africa. *Agroforestry Systems, 89*(3), 487–505. https://doi.org/10.1007/s10457-014-9780-7

Michel, D. (2023). Water and food: How, When, and why water imperils global food security. Center for strategic and international studies. https://www.csis.Org/analysis/water-and-food-how-when-and-why-water-imperils-global-food-security

Monreal, C. M., DeRosa, M., Mallubhotla, S. C., Bindraban, P. S., & Dimkpa, C. (2016). The application of nanotechnology for micronutrients in soil-plant systems. *Vitamins and Hormones, 100*, 123–157. https://doi.org/10.1016/bs.vh.2015.12.003

Nowak, B. (2021). Precision agriculture: Where do we stand? A review of the adoption of precision agriculture technologies on field crop farms in developed countries. *Agricultural Research, 10*(4), 515–522.

Perfecto, I., & Vandermeer, J. (2010). The agricultural matrix is an alternative to the land-sparing/agriculture intensification model. *Proceedings of the National Academy of Sciences, 107*(13), 5786–5791. https://doi.org/10.1073/pnas.0905455107

Pretty, J. (2008). Agricultural sustainability: Concepts, principles and evidence. *Philosophical Transactions of the Royal Society B: Biological Sciences, 363*(1491), 447–465. https://doi.org/10.1098/rstb.2007.2163

Rehman, A., Farooq, M., Lee, D. J., & Siddique, K. H. (2022). Sustainable agricultural practices for food security and ecosystem services. *Environmental Science and Pollution Research, 29*(56), 84076–84095.

Salimi, S., Almuktar, S. A., & Scholz, M. (2021). Impact of climate change on wetland ecosystems: A critical review of experimental wetlands. *Journal of Environmental Management, 286*, Article 112160.

Scherr, S. J., & McNeely, J. A. (2008). Biodiversity conservation and agricultural sustainability: Towards a new paradigm of 'ecoagriculture' landscapes. *Philosophical Transactions of the Royal Society B: Biological Sciences, 363*(1491), 477–494. https://doi.org/10.1098/rstb.2007.2165

Smith, B. D., & Anderson, A. B. (1994). Prehistoric agriculture in the Americas. *Annual Review of Anthropology, 23*(1), 247–266. https://doi.org/10.1146/annurev.anthro.23.1.247

Tilman, D., Cassman, K. G., Matson, P. A., Naylor, R., & Polasky, S. (2002). Agricultural sustainability and intensive production practices. *Nature, 418*(6898), 671–677. https://doi.org/10.1038/nature01014

Tittonell, P. (2014). Ecological intensification of agriculture—sustainable by nature. *Current Opinion in Environmental Sustainability, 8*, 53–61. https://doi.org/10.1016/j.cosust.2014.08.006

Yadav, M., Vashisht, B. B., Jalota, S. K., Jyolsna, T., Singh, S. P., Kumar, A., & Singh, G. et al. (2024). Improving water efficiencies in rural agriculture for sustainability of water resources: A review. *Water Resources Management*, 1–22.

Chapter 8
RNAi-Based Gene Silencing Technologies as New Plant Breeding Techniques (NPBT)

Ahmed Abdul Haleem Khan

Abstract With progress in time the civilization started the domestication of wild varieties of plants and farming turned source of food and feed. The subsequent growing of crop plants suffered with multiple shortcomings that need to be overcome. The tasks were resolved timely and research on contemporary crop issues brought novel varieties that withstand the resistant species with high yields. The findings of studies of genes and molecular level revolutionized the farming systems with enhanced output by using new breeds of plants that resist pathogens and pests. The reports on crop improvements are topic interest from generations that developed hybrid varieties for numerous phenomes as per the climatic conditions. There are several techniques introduced to improve crop yields in adverse habitats, i.e., classical breeding, application of induced mutations, transgenesis, gene editing, and gene silencing. The gene silencing use of RNA interference is one among the fruitful approach to develop tolerant crop varieties that effectively overcome the burden of pests and pathogens. This chapter covers the different disciplines of RNAi that contributed to get better application in field of new breeding techniques.

Keywords RNAi · Pests · Pathogens · Gene silencing · dsRNA · Breeding techniques · Crop improvement

8.1 Introduction

The domestication of wild plants for variety of purposes in different agroecosystems with high yields was target. The path of target was not easy as time to time the factors that hamper the yields like climate, pests, and pathogens dominated to reduce the yields. The farming lands were affected with global warming and climate change due to industrialization of nations, deforestation of natural habitats, lack of soil management by farming communities, and decreased or enhanced precipitation.

A. A. H. Khan (✉)
Department of Botany, Telangana University, Dichpally, Nizamabad, Telangana, India
e-mail: aahaleemkhan@gmail.com

J. Parray and A. K. Haghi (eds.), *Food and Water Security*,
https://doi.org/10.1007/978-3-032-04870-7_8

The increase in fitness for wild plants during domestication led to development to techniques that were introduced time to time resulted in cropping system well adapted to habitats different from their native substratum. The availability of wide range of plant species in the same genera with high yields and low output put forward the idea to cross between the crop varieties. The rediscovery of Mendelian concept popularized the hybridization in plants and develop improved traits in crop varieties. The classical approaches of breeding were revolutionized with information acquired from discovery of chromosomes, genes, and deoxy ribonucleic acid (DNA). Soon after the introduction of green revolution, farming systems were enabled with seeds of superior varieties to enhance the yields.

The breeding techniques were initially targeted to adapt the plants for high yields from crops and then proceeded to enhance the nutritive value in edible crops through molecular approaches (Fig. 8.1). The progress in time much focused on techniques utilizing the genetic and DNA-based traits. The techniques using ribonucleic acid in enhance anthocyanin pigment formation in petunia flower laid foundation for RNA interference (Fig. 8.2). The Nobel Prize in Physiology and Medicine of 2006, Andrew Fire and Craig Mello brought RNAi to fame that improvised gene silencing (Fig. 8.3).

RNAi is natural mechanism of gene regulation that play important part in gene silencing post transcriptionally in most of the organisms (Fig. 8.4). This technique was utilized in breeding phenomenon to improve/alter the qualitative and quantitative traits in crop plants (Fig. 8.5). The chapter hereby discusses the different perspectives of RNAi to raise the sustainable approach to apply the dsRNA and obtain the engineered plants with desired traits.

Fig. 8.1 Approaches to enhance the crop yield and quality

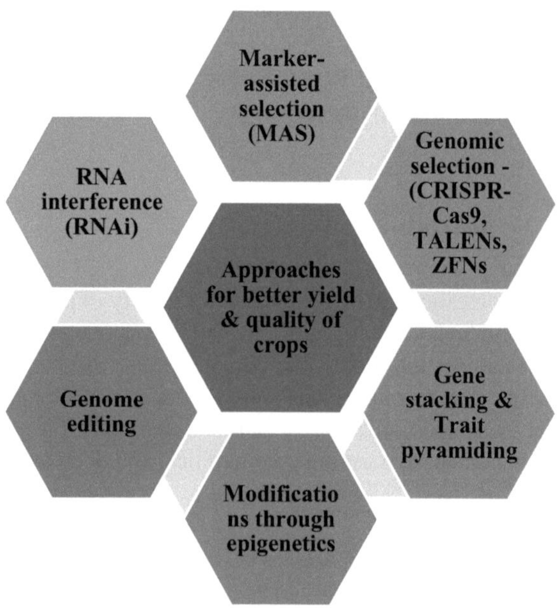

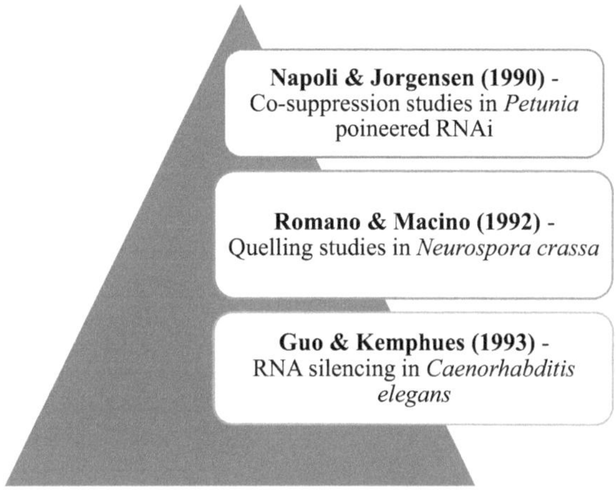

Fig. 8.2 Approaches led to development of RNA interference

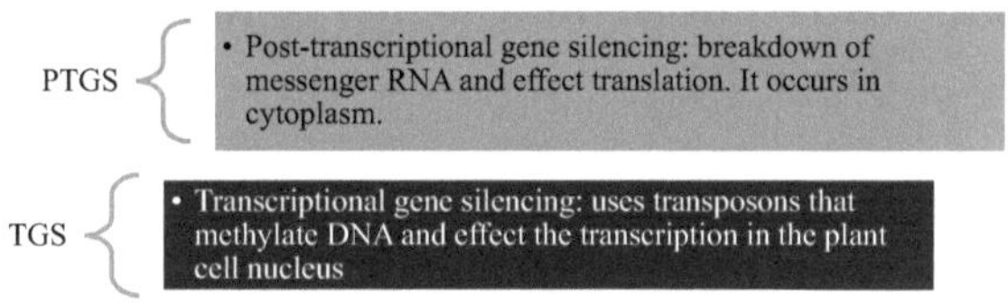

Fig. 8.3 Gene silencing processes

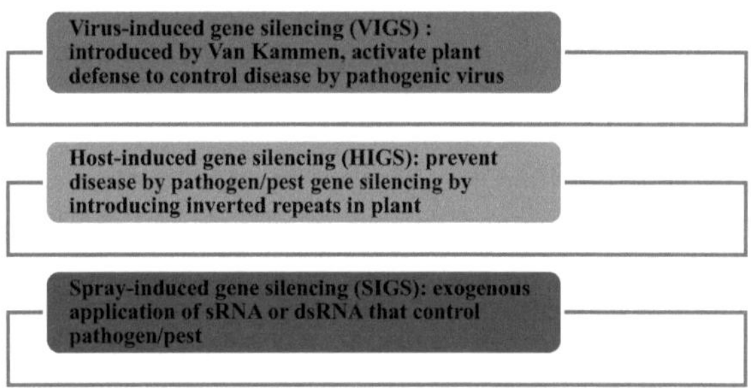

Fig. 8.4 Disease control in plants through RNA interference

The challenges for cultivated crop plants with abiotic and biotic parameters were being solved using gene-based technologies, i.e., genetic engineering that resulted in advances in information on gene functions which show significance on enzymes that effect metabolic pathways. The gene transfer either direct or using vector based were gained to develop new varieties of crop plants. The approaches using zinc finger nuclease (ZFN), transcription activator like effector nuclease (TALEN) succeeded to introduce cluster regularly interspaced short palindromic repeats (CRISPR)-based technology to target specific and manipulate the genome for improved traits in most of the organisms. The gene-based technologies gained success with unlimited drawbacks that brought the need for alternative approach. The technique of RNAi followed popularity in field of plant breeding due to its wide range of applications (Table 8.1).

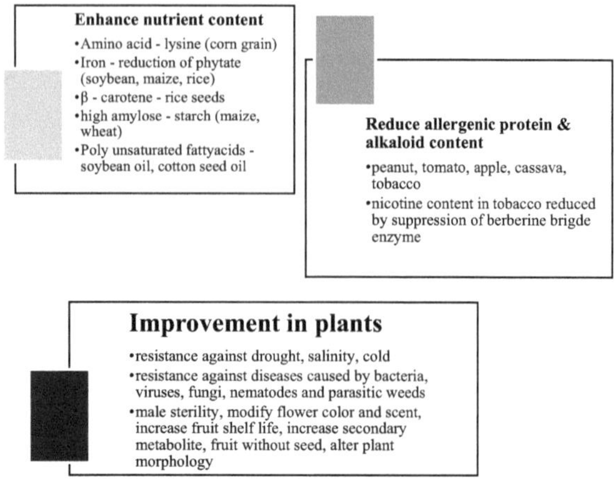

Fig. 8.5 RNAi for improved traits in edible crop plants

8.2 Different Achievements of RNAi in Field of Plant Breeding

dsRNA for manipulation of ethylene pathway

The study reported on dsRNA for its impact on seedling and fruit of tomato (*Solanum lycopersicum*) through injection and seed soaking approaches for triple response through ethylene. The findings showed dsRNA was effective in gene silencing in seedlings and fruits by regulation of ethylene pathway (Cedillo-Jimenez et al., 2024).

SiRNA for crop improvement

The gene delivery through vectors, i.e., bacteria or virus need to be updated with recent technique and the use of nanoparticle to transfer the siRNA was reported in study on *Nicotiana benthamiana.* The external application of silica nanoparticles with siRNA were effective to target different genes in the test plant. The nanoparticles employed in the study found to be better tool in gene silencing than vector-based approach (Cai et al., 2024).

Spray-induced gene silencing

The study reported on different shaped RNA-based nanoparticles on efficiency of RNAi in plant gene functions through spraying approach for dsRNA in *Panax notoginseng* and *Camellia oleifera.* The findings of square shape RNA nanoparticles synthesized using *Escherichia coli* proved effective in gene silencing. The sap sucking pests, i.e., *Aphis citricidus, A. gossypii, Myzus persicae, Melanophis sacchari, Rhopalosiphum padi,* and *Acrythosiphon pisum* were effectively controlled by spray-induced gene silencing using chemically synthesized ssRNA (Wang et al., 2024a, 2024b, 2024c).

Table 8.1 Applications of dsRNA and their prospects

Target genes	Findings	References
Nicotiana tabacum, p126 (TMV silencing suppressor) and coat protein (CP) genes	The application resulted in resistance against *Tobacco mosaic virus*	Konakalla et al. (2016)
Nicotiana benthamiana and *Vigna unguiculata* (cowpea), NIb (plants nuclear inclusion b) protein and CP (coat protein)	The aphid (*Myzus persicae*) mediated *Potyvirus bean common mosaic virus* transmission was prevented	Worrall et al., (2019)
Rice, *MoDES1*	The spray-induced dsRNA conferred resistance against fungal blast disease caused by *Magnaporthe oryzae*	Sarkar and Roy-Barman (2021)
Triticum aestivum, carboxylesterase (CbE E4) gene	The pesticide resistance by aphid (*Sitobion avenae*) reduced	Xu et al. (2014)
Arabidopsis and barley, sterol 14α-demethylase genes FgCYP51A and FgCYP51B	Growth inhibition of *Fusarium graminearum* through host-induced and spray-induced gene silencing (HIGS & SIGS)	Koch et al. (2019)
Hordeum vulgare, fungal *CYP51* genes	*Fusarium graminearum* (necrotrophic fungus) infection in test plant roots was inhibited	Biedenkopf et al. (2020)
Nicotiana tabacum, Vigna unguiculata—homologous RNA silencing	LDH (layered double hydroxide) clay nanosheets used to deliver dsRNA protected against *Pepper mild mottle virus* (PMMoV) and *Cucumber mosaic virus* (CMV)	Mitter et al. (2017)
Oryza sativa and *Zea mays,* Enhanced yellow fluorescent protein (*dsEYFP*)	Enhanced mortality of insects -*Nilaparvata lugens* (brown planthopper) and *Ostrinia furnacalis* (Asian corn borer)	Li et al. (2015)
Fusarium oxysporum and *Mycosphaerella fijiensis* banana pathogen	Essential genes for spore germination were targeted	Mumbanza et al. (2013)
Botrytis cinerea disease attenuation in Arabidopsis and tomato	Small RNAs inhibited the gray mold disease in tomato	Wang et al. (2016a, 2016b)

(continued)

Table 8.1 (continued)

Target genes	Findings	References
Protection of barley(*Hordeum vulgare* L.) from *Fusarium graminearum*	Spray on leaf with dsRNA inhibited the growth of pathogen	Koch et al. (2016)
The pathogens from oomycetes—*Phytophthora infestans* and *Hyaloperonospora parasitica;* ascomycetes -*B. cinerea, F. graminearum* and *Verticillium dahlia,* basidiomycetes—*Ustilago hordei* and *Melampsora lini* were reported for control through gene silencing	Host-induced gene silencing was found effective to control from different studies	Walker et al. (2023)

8.3 Control of Plant Pathogenic Bacteria Using dsRNA

The study of dsRNA coated with cationic poly-aspartic acid derived polymer (CPP) and layered double hydroxide (LDH) applied exogenously on *Nicotiana benthamiana, Arabidopsis thaliana* and rice plants for physiological traits and resistant against pathogenic bacteria, *Xanthomonas oryzae.* The findings of study showed improved test plant defense and resistance against bacterial pathogen (Pal et al., 2024).

8.3.1 Plant Pathogenic Fungi Control

The food crops are prone to loss by variety of pathogens and pests that attack preharvest, during standing crop and even postharvest that arise situation to protect crop plants from such harmful pathogens. The emergence of synthetic fungicides and pesticides reduced the size of infestation but the ecosystem got devasted. The integrated pest management practices were introduced to develop sustainable cropping with microbial inoculants. The use of dsRNA to control pathogen through silencing by the host defense response genes, i.e., host-induced and spray-induced gene silencing (HIGS and SIGS).

The study on use of dsRNA through spray-induced gene silencing for control of fungal plant pathogens *Aspergillus niger* (contaminate edible plant and animal products), *Botrytis cinerea* (gray mold disease), *Colletotrichum gloeosporioides* (anthracnose), *Sclerotinia sclerotiorum* (necrotrophic pathogen), *Rhizoctonia solani* (soil-borne pathogen), *Trichoderma virens,* and *Verticillium dahlia* (soil-borne pathogen) was performed by Qiao et al. (2021). The findings of study showed that the RNA uptake by fungi play role in inhibition of test fungi. There was no uptake by *Colletotrichum gloeosporioides* (pathogen), and weak uptake in a *Trichoderma virens* (beneficial) that had no suppressive effect by RNA (Qiao et al., 2021).

The study carried in vitro and whole plant approach to inhibit the rust causing fungi, i.e., *Austropuccinia psidii* and *Coleosporium plumeriae* in host plants—myrtle and frangipani by exogenous application of dsRNA spray method. The findings proved the potential of dsRNA in suppression of rust fungi (Degnan et al., 2023). The causative fungal agent (*Cercospora zeina*) gray leaf spot disease on maize was applied externally with dsRNA to study the control of pathogen. The results proved dsRNA as fungicide that can be effective against foliar pathogen (Marais et al., 2024).

Botrytis cinerea

The artificial nanovesicles were synthesized with three different formulations and used for encapsulation of dsRNA. The encapsulated RNA was applied through SIGS on grape and tomato test plants to inhibit the pathogenic fungi, i.e., *Botrytis cinerea.* The nanocarrier used were effective to increase the RNA stability and efficacy in disease control by test pathogen (Qiao et al., 2023).

Fusarium verticillioides

The fungal pathogen is major pathogen that causing huge loss in cereal crops. The seedlings of *Nicotiana benthamiana* was inoculated with *F. verticillioides* to investigate the role of RNAi in pathogenicity. The different forms of RNA were reported during the fungal pathogenesis in the host (Aydinoglu et al., 2023).

Rhizoctonia solani

The study was reported using different nanoparticle carriers (chitosan, polyethyleneimine, protamine, carbon quantum dot, polyamidoamine) for dsRNA delivery through SIGS to control rice sheath blight causative agent, i.e., *Rhizoctonia solani.* The nanocarriers were effective in improving stability of RNA and protection against the pathogen (Wang et al., 2023a, 2023b). The pathogenic fungi, i.e., *R. solani* causative agent of target spot disease in tobacco was reported to suppress through host-induced gene silencing by dsRNA exogenous application in *N. benthamiana* (Wang et al., 2024a, 2024b, 2024c). The sheath blight disease in rice was targeted through host-induced gene silencing using RNAi through tobacco rattle virus construct. The findings showed increased resistance towards pathogen, *R. solani* (Mei et al., 2024).

Sclerotinia sclerotiorum

The conventional approach to control the pathogenic fungi showed less effective to reduce the disease incidences in canola crop. The study was reported use of RNAi to treat the plants with species-specific RNA and reduce the effects of pathogens. The foliar application of dsRNA proved to control the disease symptoms by pathogenic fungi (McLoughlin et al., 2018).

The white mold, *S. sclerotiorum* infection was showed to reduce in transgenic *A. thaliana* through hairpin RNA by host-induced gene silencing approach (Walker et al., 2023). The white rot in *N. benthamiana* and *Brassica juncea* by *S. sclerotiorum* infection was reported to control through spray- and host-induced gene silencing approaches using dsRNA (Pant & Kaur, 2024).

8.3.2 Double-Stranded RNA for Plant Pathogenic Virus Control

Citrus tristeza virus

The fruits of citrus species are known serve sour taste to variety of culinary dishes with vitamin C and dietary fiber. The citrus crops are devasted by several pathogens, among them *Citrus tristeza virus*, filamentous ssRNA that restrict to phloem. This virus spreads through *Toxoptera citricidus* (insect vector) and propagation of virus infected plant material. Borah et al. (2024) reported the exogenous application of dsRNA reduced the effect of virus in test plants in vitro through different genes targeted (CP, p20, p23). The findings of the study showed the control of pathogenic virus without transgenic approach.

Maize dwarf mosaic virus

The sweet corn crop is common cultivated in both tropics and temperate regions that susceptible to wide range of infections, among them viral infection, i.e., by *Maize dwarf mosaic virus* reduce about 70% crop yield. Balassa et al. (2024) applied exogenously small interfering RNA (siRNA) on sweet corn plants to study the virus control. The findings showed that the siRNA treatment-controlled virus infection in the test plants compared to control.

Black streaked dwarf virus

The RNA interference in leaves of rice lines was studied to develop resistance against infectious virus in host plants. The findings showed effective resistance towards target pathogenic virus in genetically engineered rice with siRNA (Huang et al., 2024).

8.3.3 Double-Stranded RNA for Plant Insect Pest Control

Migratory locust (*Locusta migratoria*) and Asian corn borer (*Ostrinia furnacalis*).

The insect fluids and cultured tissues of both pests were evaluated for uptake of dsRNA and gene silencing efficiency. The results correlated the RNAi efficiency with degradation of RNA by nucleases (Fan et al., 2022).

Migratory locust (*Locusta migratoria*), American cockroach (*Periplaneta americana*), Tobacco caterpillar (*Spodoptera litura*), Tenebrionid beetle (*Zophobas atratus*).

The four insect species were investigated for sensitivity to RNAi, by administration of dsRNA. The results showed RNAi was effective to degrade the secretion of midgut that sensitize the test species (Wang et al., 2016a, 2016b).

Tobacco cutworm (*Spodoptera litura*)

The cutworm was investigated for dsRNase responsible for degradation of RNA that leads to RNAi efficiency. The larval stages were reported to show high dsRNA degradation and reduced efficiency of RNAi (Peng et al., 2020).

Melon fly (*Zeugodacus cucurbitae*)

The edible fruit yielding cucurbit crops were known to be infested by insect pest, *Zeugodacus cucurbitae* that results in wide range of crop losses. The pest lay eggs on fruit of crop plants and larval forms feed the pulp. Ahmad et al. (2024) investigated the use of RNA degrading enzymes to target the insect in species-specific approach through RNA interference (RNAi) to avoid the crop damage by the pest. The double-stranded RNases were designed to specific gene that target the coatamer subunit of insect pest. The investigators used two dsRNases that resulted in larval mortality in efficient manner. The study showed efficacy in control of pest safely than the recalcitrant synthetic pesticides.

Mediterranean (med) fly (*Ceratitis capitata*)

The wide range of agri-crops yielding fruits and vegetables are attacked by medfly with huge loss economically. There are multiple approaches to reduce the burden of pest species using conventional chemical agents and the eco-safe sustainable ways are need of farming systems. Ortola et al. (2024) reported in vitro study using dsRNA from *Escherichia coli* to target seven target genes in the medfly. The results proved efficient control of fly through oral transfer of dsRNA as potential avenue to protect crop plants from pest.

White-backed planthopper (*Sogatella furcifera*)

The study was reported on biopesticide that targeted pest i.e. white backed planthopper by employing spray-induced gene silencing through nanocarrier. The findings showed silencing mechanism based on post-transcription that affected the growth and development of pest (Ma et al., 2024).

Chickpea pod borer (*Helicoverpa armigera*)

The increased tolerance towards conventional approaches, i.e., synthetic insecticides and transgenic plant varieties by pod borer switched to exogenous application of dsRNA was effective in chickpea pod damage control with no effect on non-target insect species (Kolge et al., 2021). The cotton bollworm (*H. armigera*) was investigated to its control studies through use of transgenic *Nicotiana benthamiana* that express anti-insect RNA. The insects feeding leaf of this plants were found with abnormalities and altered growth resulted into mortality. The results were novel in control of bollworm through herbivory (Bally et al., 2020).

Leafhoppers (*Euscelidus variegatus, Scaphoideus titanus*)

The leafhoppers are well known vectors of obligate intracellular pathogenic bacteria, i.e., phytoplasma that colonize the host plant phloem tissues and result in disease condition. The study reported on use of dsRNA mediated gene silencing proved effective in insect control (Rossi et al., 2024).

Aphid (*Aphis gossypii*)

The aphids from cotton field were studied for control of phloem sucking behavior by using dsRNA. The findings showed the sucrase gene was targeted through RNAi against aphid (Uma et al., 2023).

Aphid (*Sitobion avenae*)

The wheat crop reported for huge losses by infestation by *S. avenae* that was target in study through RNAi to control the pest. The results showed effective control of pest via target of voltage gated sodium channels. The aphid was targeted laccase in grain aphid through RNAi. The transgenic wheat of lac gene showed pest mortality effectively without negative effects on non-target (Rafique et al., 2023).

Fire ant (*Solenopsis invicta*)

The insect pest was studied for its control through oral administration of dsRNA with liposomes. The results showed efficient in mortality through RNAi and laid path for drug development to target the nucleic acid (Wang et al., 2024a, 2024b, 2024c).

Fall armyworm (*Spodoptera frugiperda*)

The pest *S. frugiperda* is known notorious in agriculture that results in huge crop losses. The study explored the use of nanoliposomes to pass the dsRNA for methoprene tolerance and control the pest population. The nanoparticle used for application of RNA proved efficient in interference and reduce the pest incidence (Su et al. 2023). The study was carried using cationic biopolymer, protamine sulfate and cellfectin as formulation to transfer the dsRNA and control of *S. frugiperda.* The dsRNA passed into the pest was found stable and effective in control (Dhandapani et al., 2022).

Asian citrus psyllid (*Diaphorina citri*)

The pest psyllid found popular as vector of pathogenic bacteria, *Candidatus* sp., that cause disease in citrus plants. The use of dsRNA for control of psyllid was reported to increase the mortality and found effective in pest management in citrus crops (Galdeano et al., 2017).

Butterfly (*Papilio xuthus*)

The report on insect model, butterfly *P. xuthus,* for RNA degrading enzyme, dsRNase showed genes knock down and gene silencing through RNAi. The results concluded improved efficiency of RNAi through dsRNase genes (Shu et al., 2024).

Sugarcane weevil (*Sphenophorus levis*)

The plantations of sugarcane crop were known to infest by weevil that huge losses commercially. The control studies were reported using dsRNA bioassay through injection in weevil larvae showed effective down regulation of genes without negative effect on plant (Mohan et al., 2021).

Cabbage stem flea beetle (*Psylliodes chrysocephala*)

The test beetle was reported for resistance to synthetic insecticide, pyrethroid and studied for their control using RNAi. The oral delivery of dsRNA in the pest showed high mortality without negative effect on non-target insect species (Cedden et al., 2024).

Australian sheep blowfly (*Lucilia cuprina*)

The blowfly poses major threat sheep animal-based industry due to economic losses. The synthetic insecticide resistance by pest population triggered alternative approaches as RNAi. The study using cell lines from embryo of blowfly were employed that proved efficiency of dsRNA in control of pest (Yang et al., 2024).

Asian long-horned beetle (*Anoplophora glabripennis*)

The study reported on control of beetle through dsRNA to target apoptosis inhibition. The larva of ectoparasitoid was employed to locate the beetle and the RNA showed no effect on parasitoid. The beetle was effectively controlled by RNAi (Wang et al., 2023a, 2023b). The beetle odorant binding proteins were targeted to pest control. The RNAi revealed showed potential to control the pest efficiently through silencing of odorant-related genes (Zhang et al., 2024). The study used bioassay by feeding for oral delivery of dsRNA through sucrose in test beetle was proved effective in mortality of larvae and RNAi was novel approach in pest control (Dhandapani et al., 2020).

Sweetpotato weevil (*Cylas puncticollis*)

The study was reported to control the weevil through RNAi using dsRNA via injection and oral delivery. The injection mode was effective in gene silencing and control of weevil (Prentice et al., 2019).

Asian corn borer (*Ostrinia furnacalis*)

The corn borer was investigated for the knock down of RNAi related nuclease that enhanced efficiency of RNAi. The findings proved the targeted nuclease that increased small RNAs (Guan et al. 2018). The corn borer was studied for four different dsRNase genes and found the selected genes were active in different stages of insect. Among the genes, OfdsRNase2 was responsible for decrease in RNAi efficiency.

European corn borer (*O. nubilalis*)

The RNAi efficiency was tested in *O. nubilalis* by injection and oral feeding of dsRNA. The dsRNA was rapidly degraded in the tissues of insect that was suggested for poor efficiency of RNAi (Cooper et al., 2020).

Corn earworm (*Helicoverpa zea*)

The study was reported by three types of delivery of dsRNA, i.e., eggs (microinjection and soaking) and larval feeding to induce the effects of RNAi in insect. The microinjection was effective in silencing than other approaches of delivery as per the findings (Wang et al., 2018).

Red flour beetle (*Tribolium castaneum*) and pea aphid (*Acyrthosiphon pisum*)

The study was carried for gene silencing in beetle and aphid by feeding dsRNA using peptide nanomaterial. The ingestion of material accompanied with dsRNA was effective in killing of both the insects (Avila et al., 2018).

Rice striped stem borer (*Chilo suppressalis*)

The study was reported three different nanoparticles to encapsulate the dsRNA and evaluated the effect of RNA bioactivity in test insect. The carbon nano-dots were among the effect mediator that resulted gene silencing efficiently than other two nanomaterials i.e. chitosan, lipofectamine (Wang et al., 2020).

Potato beetles (*Leptinotarsa decemlineata*)

This insect is popular in potato growing areas of world as major pest that target the foliage part of crop with huge losses to farming community. The gene silencing in pest population was reported using dsRNA and RNAi using S-adenosyl-L-homocysteine hydrolase as target in first to fourth stage larvae. The findings showed mortality of larval forms based on concentration dependently as efficient approach to control pest through RNAi (Guo et al., 2015). The study reported on two nuclease genes in gut of test beetle population for sensitivity of dsRNA. The results showed effective control of insect in potato plants protection through RNAi (Spit et al., 2017). The studies on test insect targeting four genes (dsActin, dsHSP70, dsDynamin, and dsIAP) through dsRNA in both larva and adult were proved alternative chemical insecticide that resulted in gene silencing with enhanced mortality (Maximo et al., 2020).

Spotted spider mite (*Tetranychus urticae*)

The study on control of *T. urticae* in apple tree plantation was reported using dsRNA for RNAi based gene silencing. The tree trunk was injected with to deliver the dsRNA and found results were effective for abundance of dsRNA in treated plants (Wise et al., 2022).

8.4 Conclusion

The cultivation of crop plants in the current century is backbone of economy around the globe with several challenges. The introduction of superior varieties of crop seeds with addition of modern techniques is ever lasting to chase the challenges. The gene silencing with use of RNAi technique turned emerging approach in modern times that solved numerous problems of crop improvement. The use of RNAi overcome the shortcomings of transgenic plants in upbringing the modified traits in plants with desired qualities.

References

Ahmad, S., Jamil, M., Jaworski, C. C., & Luo, Y. (2024). Double-stranded RNA degrading nuclease affects RNAi efficiency in the melon fly. *Zeugodacus Cucurbitae. J Pest Sci., 97*, 397–409. https://doi.org/10.1007/s10340-023-01637-1

Avila, L. A., Chandrasekar, R., Wilkinson, K. E., Balthazor, J., Heerman, M., Bechard, J., Brown, S., Park, Y., Dhar, S., Reeck, G. R., & Tomich, J. M. (2018). Delivery of lethal dsRNAs in insect diets by branched amphiphilic peptide capsules. *Journal of Controlled Release, 273*, 139–146. https://doi.org/10.1016/j.jconrel.2018.01.010

Aydinoglu, F., & Kuloglu, A. (2023). *Nicotiana benthamiana* as a model plant host for *Fusarium verticillioides* to investigate RNA interference, cross-kingdom RNA exchange, and competitive endogenous RNA (ceRNA) network. *Molecular Biology Reports, 50*(10), 8061–8072. https://doi.org/10.1007/s11033-023-08698-6

Balassa, K., Balassa, G., Almási, A., Visnovitz, T., & Rudnoy, S. (2024). Detection of exogenous siRNA inside sweet corn bundle sheath cells and the RNAi dynamics in the early stage of *Maize dwarf mosaic virus* infection. *Physiology and Molecular Biology of Plants, 30*(8), 1265–1276. https://doi.org/10.1007/s12298-024-01500-2

Bally, J., Fishilevich, E., Doran, R. L., Lee, K., de Campos, S. B., German, M. A., Narva, K. E., & Waterhouse, P. M. (2020). Plin-amiR, a pre-microRNA-based technology for controlling herbivorous insect pests. *Plant Biotechnology Journal, 18*(9), 1925–1932. https://doi.org/10.1111/pbi.13352

Bally, J., Fishilevich, E., Bowling, A. J., Pence, H. E., Narva, K. E., & Waterhouse, P. M. (2018). Improved insect-proofing: Expressing double-stranded RNA in chloroplasts. *Pest Management Science, 74*(8), 1751–1758. https://doi.org/10.1002/ps.4870

Biedenkopf, D., Will, T., Knauer, T., Furch, A. C. H., Busche, T., & Koch, A. (2020). Systemic spreading of exogenous applied RNA biopesticides in the crop plant *Hordeum vulgare. ExRNA, 2*, 12. https://doi.org/10.1186/s41544-020-00052-3

Borah, M., Nath, P. D., Chaudhury, S. P., Biswas, K. K., Patil, B. L., & Voloudakis, A. (2024). Topical application of dsRNAs targeting *Citrus tristeza virus* (CTV) reduces its titer in the CTV infected sweet orange (*Citrus sinensis*). *European Journal of Plant Pathology, 168*, 273–278. https://doi.org/10.1007/s10658-023-02752-6

Cai, Y., Liu, Z., Wang, H., Meng, H., & Cao, Y. (2024). Mesoporous silica nanoparticles mediate SiRNA delivery for long-term multi-gene silencing in intact plants. *Advancement of Science, 11*, 2301358. https://doi.org/10.1002/advs.202301358

Chen, A., Halilovic, L., Shay, J. H., Koch, A., Mitter, N., & Jin, H. (2023). Improving RNA-based crop protection through nanotechnology and insights from cross-kingdom RNA trafficking. *Current Opinion in Plant Biology, 76*, Article 102441. https://doi.org/10.1016/j.pbi.2023.102441

Cedillo Jimenez, C. A., Guevara-Gonzalez, R. G., & Cruz-Hernandez, A. (2024). Exogenous dsRNA sequence based on miR1917 downregulates its target gene related to ethylene signaling in tomato seedlings and fruit. *Scientia Horticulturae., 331*, Article 113090. https://doi.org/10.1016/j.scienta.2024.113090

Cedden, D., Güney, G., Debaisieux, X., Scholten, S., Rostás, M., & Bucher, G. (2024). Effective target genes for RNA interference-based management of the cabbage stem flea beetle. *Insect Molecular Biology*. https://doi.org/10.1111/imb.12942

Cooper, A. M. W., Yu, Z., Biondi, M., Song, H., Silver, K., Zhang, J., & Zhu, K. Y. (2020). Stability of double-stranded RNA in gut contents and hemolymph of *Ostrinia nubilalis* larvae. *Pesticide Biochemistry and Physiology, 169*, Article 104672. https://doi.org/10.1016/j.pestbp.2020.104672

Dhandapani, R. K., Gurusamy, D., & Palli, S. R. (2022). Protamine-Lipid-dsRNA nanoparticles improve RNAi efficiency in the fall armyworm. *Spodoptera Frugiperda. J Agric Food Chem., 70*(22), 6634–6643. https://doi.org/10.1021/acs.jafc.2c00901

Dhandapani, R. K., Duan, J. J., & Palli, S. R. (2020). Orally delivered dsRNA induces knockdown of target genes and mortality in the Asian long-horned beetle. *Anoplophora Glabripennis. Arch Insect Biochem Physiol., 104*(4), Article e21679. https://doi.org/10.1002/arch.21679

Degnan, R. M., McTaggart, A. R., Shuey, L. S., Pame, L. J. S., Smith, G. R., Gardiner, D. M., Nock, V., Soffe, R., Sale, S., Garrill, A., Carroll, B. J., Mitter, N., & Sawyer, A. (2023). Exogenous double-stranded RNA inhibits the infection physiology of rust fungi to reduce symptoms in planta. *Molecular Plant Pathology, 24*(3), 191–207. https://doi.org/10.1111/mpp.13286

Fan, Y., Song, H., Abbas, M., Wang, Y., Liu, X., Li, T., Ma, E., Zhu, K. Y., & Zhang, J. (2022). The stability and sequence cleavage preference of dsRNA are key factors differentiating RNAi efficiency between migratory locust and Asian corn borer. *Insect Biochemistry and Molecular Biology, 143*, Article 103738. https://doi.org/10.1016/j.ibmb.2022.103738

Galdeano, D.M., Breton, M.C., Lopes, J.R., Falk, B.W., Machado, M.A., (2017) Oral delivery of double-stranded RNAs induces mortality in nymphs and adults of the Asian citrus psyllid, *Diaphorina citri. PLoS One. 12*(3), e0171847. https://doi.org/10.1371/journal.pone.0171847

Guo, W. C., Fu, K. Y., Yang, S., Li, X. X., & Li, G. Q. (2015). Instar-dependent systemic RNA interference response in *Leptinotarsa decemlineata* larvae. *Pesticide Biochemistry and Physiology, 123*, 64–73. https://doi.org/10.1016/j.pestbp.2015.03.006

Huang, C., Wang, Z., Zhu, P., Xu, W., Du, Z., Qian, Y., & Fu, W. (2024). RNAi-based genetically engineered rice resistant to black-streaked dwarf virus does not show adverse genetic effects: A multi-omics analysis. *Plants People Planet., 6*, 622–639. https://doi.org/10.1002/ppp3.10477

Koch, A., Hofle, L., Werner, B. T., Imani, J., Schmidt, A., Jelonek, L., & Kogel, K. H. (2019). SIGS vs HIGS: A study on the efficacy of two dsRNA delivery strategies to silence Fusarium FgCYP51 genes in infected host and non-host plants. *Molecular Plant Pathology, 20*(12), 1636–1644. https://doi.org/10.1111/mpp.12866

Konakalla, N. C., Kaldis, A., Berbati, M., Masarapu, H., & Voloudakis, A. E. (2016). Exogenous application of double-stranded RNA molecules from TMV p126 and CP genes confers resistance against TMV in tobacco. *Planta, 244*(4), 961–969. https://doi.org/10.1007/s00425-016-2567-6

Kolge, H., Kadam, K., Galande, S., Lanjekar, V., & Ghormade, V. (2021). New frontiers in pest control: Chitosan nanoparticles-shielded dsRNA as an effective topical RNAi spray for gram podborer biocontrol. *ACS Applied Bio Materials, 4*(6), 5145–5157. https://doi.org/10.1021/acsabm.1c00349

Koch, A., Biedenkopf, D., Furch, A., Weber, L., Rossbach, O., Abdellatef, E., Linicus, L., Johannsmeier, J., Jelonek, L., Goesmann, A., Cardoza, V., McMillan, J., Mentzel, T., & Kogel, K. H. (2016). An RNAi-based control of *Fusarium graminearum* infections through spraying of long dsRNAs involves a plant passage and is controlled by the fungal silencing machinery. *PLoS Pathogens, 12*, 1–22. https://doi.org/10.1371/journal.ppat.1005901

Li, H., Guan, R., Guo, H., & Miao, X. (2015). New insights into an RNAi approach for plant defence against piercing-sucking and stem-borer insect pests. *Plant, Cell and Environment, 38*(11), 2277–2285. https://doi.org/10.1111/pce.12546

Marais, I., Buitendag, C., Duong, T. A., Crampton, B. G., Theron, J., Kidanemariam, D., & Berger, D. K. (2024). Double-stranded RNA uptake for the control of the maize pathogen *Cercospora zeina. Plant Pathology., 73*, 1480–1490. https://doi.org/10.1111/ppa.13909

Maximo, W.P.F., Howell, J.L., Mogilicherla, K., Basij, M., Chereddy, S.C.R.R., Palli, S.R. (2020). Inhibitor of apoptosis is an effective target gene for RNAi-mediated control of Colorado potato beetle, *Leptinotarsa decemlineata. Arch Insect Biochem Physiol. 104*(4), e21685. https://doi.org/10.1002/arch.21685

McLoughlin, A.G., Wytinck, N., Walker, P.L., Girard, I.J., Rashid, K.Y., de Kievit, T., Fernando, W.G.D., Whyard, S., Belmonte, M.F., (2018) Identification and application of exogenous dsRNA confers plant protection against *Sclerotinia sclerotiorum* and *Botrytis cinerea. Sci Rep. 9;8*(1), 7320. https://doi.org/10.1038/s41598-018-25434-4

Ma, Y. F., Liu, T. T., Zhao, Y. Q., Luo, J., Feng, H. Y., Zhou, Y. Y., Gong, L. L., Zhang, M. Q., He, Y. Y., Hull, J. J., Dewer, Y., He, M., & He, P. (2024). SRNA interference-screening of potentially lethal gene targets in the white-backed planthopper *Sogatella furcifera* via a spray-induced and nanocarrier delivered gene silencing system. *Journal of Agriculture and Food Chemistry, 72*(2), 1007–1016. https://doi.org/10.1021/acs.jafc.3c05659

Mei, Z., Xiaoxue, L., Jun, W., Erxun, Z., & Canwei, S. (2024). Host-induced gene silencing of effector AGLIP1 enhanced resistance of rice to *Rhizoctonia solani* AG1-IA. *Rice Science., 31*(4), 463–474. https://doi.org/10.1016/j.rsci.2024.04.005

Mitter, N., Worrall, E. A., Robinson, K. E., Li, P., Jain, R. G., Taochy, C., Fletcher, S. J., Carroll, B. J., Lu, G. Q., & Xu, Z. P. (2017). Clay nanosheets for topical delivery of RNAi for sustained protection against plant viruses. *Nat Plants., 3*, 16207. https://doi.org/10.1038/nplants.2016.207

Mohan, C., Shibao, P. Y. T., de Paula, F. F. P., Toyama, D., Vieira, M. A. S., Figueira, A., Scotton, D., Soares-Costa, A., & Henrique-Silva, F. (2021). HRNAi-mediated knock-down of *Sphenophorus levis* V-ATPase E in transgenic sugarcane (*Saccharum* spp interspecific hybrid) affects the insect growth and survival. *Plant Cell Reports, 40*(3), 507–516. https://doi.org/10.1007/s00299-020-02646-5

Mumbanza, F.M., Kiggundu, A., Tusiime, G., Tushemereirwe, W.K., Niblett, C., Bailey, A. (2013) In vitro antifungal activity of synthetic dsRNA molecules against two pathogens of banana, *Fusarium oxysporum* f. sp. *cubense* and *Mycosphaerella fijiensis. Pest Manag Sci.* 69(10):1155–1162. https://doi.org/10.1002/ps.3480

Ortola, B., Urbaneja, A., Eiras, M., Perez-Hedo, M., & Daros, J. A. (2024). RNAi-mediated silencing of Mediterranean fruit fly (*Ceratitis capitata*) endogenous genes using orally-supplied double-stranded RNAs produced in Escherichia coli. *Pest Management Science, 80*(3), 1087–1098. https://doi.org/10.1002/ps.7839

Pal, G., Ingole, K. D., Yavvari, P. S., Verma, P., Kumari, A., Chauhan, C., Chaudhary, D., Srivastava, A., Bajaj, A., & Vemanna, R. S. (2024). Exogenous application of nanocarrier-mediated double-stranded RNA manipulates physiological traits and defence response against bacterial diseases. *Molecular Plant Pathology, 25*(1), Article e13417. https://doi.org/10.1111/mpp.13417

Pant, P., & Kaur, J. (2024). Control of *Sclerotinia sclerotiorum* via an RNA interference (RNAi)-mediated targeting of SsPac1 and SsSmk1. *Planta, 259*(6), 153. https://doi.org/10.1007/s00425-024-04430-1

Peng, Y., Wang, K., Zhu, G., Han, Q., Chen, J., Elzaki, M. E. A., Sheng, C., Zhao, C., Palli, S. R., & Han, Z. (2020). Identification and characterization of multiple dsRNases from a lepidopteran insect, the tobacco cutworm, *Spodoptera litura* (Lepidoptera: Noctuidae). *Pesticide Biochemistry and Physiology, 162*, 86–95. https://doi.org/10.1016/j.pestbp.2019.09.011

Prentice, K., Smagghe, G., Gheysen, G., & Christiaens, O. (2019). Nuclease activity decreases the RNAi response in the sweetpotato weevil *Cylas puncticollis. Insect Biochemistry and Molecular Biology, 110*, 80–89. https://doi.org/10.1016/j.ibmb.2019.04.001

Qiao, L., Nino-Sanchez, J., Hamby, R., Capriotti, L., Chen, A., Mezzetti, B., & Jin, H. (2023). Artificial nanovesicles for dsRNA delivery in spray-induced gene silencing for crop protection. *Plant Biotechnology Journal, 21*(4), 854–865. https://doi.org/10.1111/pbi.14001

Qiao, L., Lan, C., Capriotti, L., Ah-Fong, A., Nino Sanchez, J., Hamby, R., Heller, J., Zhao, H., Glass, N. L., Judelson, H. S., Mezzetti, B., Niu, D., & Jin, H. (2021). Spray-induced gene silencing for disease control is dependent on the efficiency of pathogen RNA uptake. *Plant Biotechnology Journal, 19*(9), 1756–1768. https://doi.org/10.1111/pbi.13589

Rafique, A., Afroz, A., Zeeshan, N., Rashid, U., Khan, M. A. U., Irfan, M., Chatha, W., Khan, M. R., & Rehman, N. (2023). Production of *Sitobion avenae*-resistant *Triticum aestivum* cvs using laccase as RNAi target and its systemic movement in wheat post dsRNA spray. *PLoS ONE, 18*(5), Article e0284888. https://doi.org/10.1371/journal.pone.0284888

Rossi, M., Ottati, S., Bucci, L., Fusco, A., Abba, S., Bosco, D., Marzachi, C., & Galetto, L. (2024). Lab-scale method for plant-mediated delivery of dsRNAs to phloem-feeding leafhoppers. *Journal of Pest Science, 97*, 455–467. https://doi.org/10.1007/s10340-023-01670-0

Sarkar, A., & Roy-Barman, S. (2021). Spray-induced silencing of pathogenicity gene *MoDES1 via* exogenous double-stranded RNA can confer partial resistance against fungal blast in rice. *Frontiers in Plant Science, 12*, Article 733129. https://doi.org/10.3389/fpls.2021.733129

Spit, J., Philips, A., Wynant, N., Santos, D., Plaetinck, G., Vanden Broeck, J. (2017). Knockdown of nuclease activity in the gut enhances RNAi efficiency in the Colorado potato beetle, *Leptinotarsa decemlineata*, but not in the desert locust, *Schistocerca gregaria. Insect Biochem Mol Biol.* 81, 103–116. https://doi.org/10.1016/j.ibmb.2017.01.004

Shu, Q., Liu, G. C., He, J. W., Hu, P., Dong, Z. W., Zhao, R. P., Zhang, H. R., & Li, X. Y. (2024). RNAi efficiency is enhanced through knockdown of double-stranded RNA-degrading enzymes in butterfly *Papilio xuthus. Archives of Insect Biochemistry and Physiology, 115*(4), Article e22113. https://doi.org/10.1002/arch.22113

Su, C., Liu, S., Sun, M., Yu, Q., Li, C., Graham, R. I., Wang, X., Wang, X., Xu, P., & Ren, G. (2023). Delivery of Methoprene-tolerant dsRNA to improve RNAi efficiency by modified liposomes for pest control. *ACS Applied Materials & Interfaces, 15*(10), 13576–13588. https://doi.org/10.1021/acsami.2c20151

Uma, G. S., Saakre, M., Singh, J., & Kalia, V. K. (2023). Double-stranded RNA mediated knockdown of sucrase gene induced mortality and reduced offspring production in *Aphis gossypii. Functional & Integrative Genomics, 23*(4), 305. https://doi.org/10.1007/s10142-023-01233-7

Wang, J. D., Chen, Y. H., Zhang, Y. X., Lin, J. W., Gao, S. J., Tang, B. Z., & Hou, Y. M. (2024a). Establishment of RNAi-mediated pest control method for red imported fire ant. *Solenopsis Invicta. J Agric Food Chem., 72*(19), 10936–10943. https://doi.org/10.1021/acs.jafc.4c00654

Wang, Y., Guo, Y., Guo, S., Qi, L., Li, B., Jiang, L., Xu, C., An, M., & Wu, Y. (2024b). RNA interference-based exogenous double-stranded RNAs confer resistance to *Rhizoctonia solani* AG-3 on *Nicotiana tabacum. Pest Management Science, 80*(4), 2170–2178. https://doi.org/10.1002/ps.7962

Wang, X., Faucher, J., Dhandapani, R. K., Duan, J. J., & Palli, S. R. (2023a). Potential effects of RNA interference of Asian longhorned beetle on its parasitoid. *Pest Management Science, 79*(4), 1557–1565. https://doi.org/10.1002/ps.7328

Wang, Y., Li, X., Zhu, C., Yi, S., Zhang, Y., & Hong, Z. (2024c). Plant-derived artificial miRNA effectively reduced the proliferation of aphid (Aphidoidea) through spray-induced gene silencing. *Pest Management Science, 80*(9), 4322–4332. https://doi.org/10.1002/ps.8138

Wang, Y., Yan, Q., Lan, C., Tang, T., Wang, K., Shen, J., & Niu, D. (2023b). Nanoparticle carriers enhance RNA stability and uptake efficiency and prolong the protection against Rhizoctonia solani. *Phytopathol Res, 5*, 2. https://doi.org/10.1186/s42483-023-00157-1

Wang, M., Weiberg, A., Lin, F. M., Thomma, B. P., Huang, H. D., & Jin, H. (2016a). Bidirectional cross-kingdom RNAi and fungal uptake of external RNAs confer plant protection. *Nat Plants., 2*, 16151. https://doi.org/10.1038/nplants.2016.151

Walker, P. L., Ziegler, D. J., Giesbrecht, S., McLoughlin, A., Wan, J., Khan, D., Hoi, V., Whyard, S., & Belmonte, M. F. (2023). Control of white mold (*Sclerotinia sclerotiorum*) through plant-mediated RNA interference. *Science and Reports, 13*(1), 6477. https://doi.org/10.1038/s41598-023-33335-4

Wang, K., Peng, Y., Chen, J., Peng, Y., Wang, X., Shen, Z., & Han, Z. (2020). Comparison of efficacy of RNAi mediated by various nanoparticles in the rice striped stem borer (*Chilo suppressalis*). *Pesticide Biochemistry and Physiology, 165*, Article 104467. https://doi.org/10.1016/j.pestbp.2019.10.005

Wang, J., Gu, L., Knipple, D. C. (2018). Evaluation of some potential target genes and methods for RNAi-mediated pest control of the corn earworm. *Helicoverpa zea. Pestic Biochem Physiol.* 149, 67–72. https://doi.org/10.1016/j.pestbp.2018.05.012

Wang, K., Peng, Y., Pu, J., Fu, W., Wang, J., & Han, Z. (2016b). Variation in RNAi efficacy among insect species is attributable to dsRNA degradation in vivo. *Insect Biochemistry and Molecular Biology, 77*, 1–9. https://doi.org/10.1016/j.ibmb.2016.07.007

Wise, J. C., Wise, A. G., Rakotondravelo, M., Vandervoort, C., Seeve, C., & Fabbri, B. (2022). Trunk injection delivery of dsRNA for RNAi-based pest control in apple trees. *Pest Management Science, 78*(8), 3528–3533. https://doi.org/10.1002/ps.6993

Worrall, E. A., Bravo-Cazar, A., Nilon, A. T., Fletcher, S. J., Robinson, K. E., Carr, J. P., & Mitter, N. (2019). Exogenous application of RNAi-inducing double-stranded RNA inhibits aphid-mediated transmission of a plant virus. *Frontiers in Plant Science, 10*, 265. https://doi.org/10.3389/fpls.2019.00265

Xu, L., Duan, X., Lv, Y., Zhang, X., Nie, Z., Xie, C., Ni, Z., & Liang, R. (2014). Silencing of an aphid carboxylesterase gene by use of plant-mediated RNAi impairs *Sitobion avenae* tolerance of Phoxim insecticides. *Transgenic Research, 23*(2), 389–396. https://doi.org/10.1007/s11248-013-9765-9

Yang, Y., Zhang, B., Yong, J., James, P., Xu, Z. P., Mitter, N., Mahony, T. J., & Mody, K. T. (2024). The use of cell and larval assays to identify target genes for RNA interference-meditated control of the Australian sheep blowfly (*Lucilia cuprina*). *Pest Management Science, 80*(9), 4686–4698. https://doi.org/10.1002/ps.8190

Zhang, S., Li, M., Zhao, Y., Niu, Y., Liu, C., Tao, J., & Zong, S. (2024). Silencing the odorant co-receptor (Orco) in *Anoplophora glabripennis* disrupts responses to pheromones and host volatiles. *Pesticide Biochemistry and Physiology, 203*, Article 105968. https://doi.org/10.1016/j.pestbp.2024.105968

Chapter 9
Food Security and Sustainability Through Energy-Efficient Convective Drying

G. M. Chaitradeepa, Vinayak Deshi, Pooja Eranna, K. N. Hanumantharaju, and Chennappa Gurikar

Abstract As the world's population continues to expand, the necessity to ensure food security and sustainability becomes crucial. In order to combat hunger and malnutrition, food security ensures that all individuals have access to wholesome food. By ensuring sustainable food production methods uphold long-term profitability, safeguard the environment, and conserve resources. Drying prolongs the shelf life of perishable goods, minimizes waste, and prevents spoiling all of which are critical for food security. Convective drying, which takes place in the dryer at various temperatures, shortens the drying time as the temperature of the drying air rises for every sample. It guarantees a consistent and dependable food supply, particularly in areas with seasonal fluctuations or restricted availability of fresh crops. In convective drying, hot and dry gas (usually air) is used to remove water vapor from the food's surface and to provide the heat required for evaporation. These efficient drying techniques address the problem of feeding a growing global population, mitigate the effects of food shortages, and strengthen the food supply chain. The importance of convective drying on prolonged shelf life of food products, minimizing losses and encouraging sustainable agricultural practices needs to be addressed for the sustainability. This chapter contributes valuable insights on drying principle, various drying methods, importance and applications ensuring the food systems resilience and fostering a more secure and sustainable future.

Keywords Shelf life · Sustainability · Convective drying · Food security · Food chain · Agricultural practices

G. M. Chaitradeepa · P. Eranna · K. N. Hanumantharaju (✉) · C. Gurikar
Department of Food Technology, FLAHS, Ramaiah University of Applied Sciences, Bangalore, Karnataka, India
e-mail: hanumantharaju.ft.ls@msruas.ac.in

V. Deshi
Department of Food Science and Postharvest Technology, Bihar Agricultural University, Sabour, Bhagalpur, Bihar, India

J. Parray and A. K. Haghi (eds.), *Food and Water Security*,
https://doi.org/10.1007/978-3-032-04870-7_9

9.1 Introduction

Understanding Food Security and Sustainability is crucial for ensuring a stable food supply and minimizing environmental impact (Wang et al., 2021). It involves implementing efficient and innovative processes in the food production and preservation industry to maintain food safety, stability, and nutritional value. Food security refers to the availability, access, and utilization of food by individuals or populations (García-Diez et al., 2021). Food security also encompasses the concept of sustainability, which involves meeting present food needs without compromising the ability of future generations to meet their own needs (Clapp et al., 2022). By utilizing energy-efficient convective drying techniques, we can reduce food waste and improve food security and sustainability. During food production, transport, storage, and final consumption, various preservation techniques can be employed to ensure food safety and stability. These techniques include pasteurization, sterilization, cooling, freezing, ohmic heating, microwave, and radio frequency.

To achieve this, suitable, effective, and economic methods such as convective drying can be utilized (Sahoo et al., 2022). Convective drying, as a method of food preservation, plays a significant role in achieving food security and sustainability. By using convective drying, we can extend the shelf life of food, reduce food waste, and minimize reliance on non-renewable energy sources (Fathi et al., 2022). Furthermore, convective drying offers several advantages in terms of energy efficiency. It allows for the removal of moisture from food products using hot air or gases, reducing the need for excessive energy consumption. By utilizing convective drying, food can be preserved and dehydrated more efficiently, especially in regions with limited access to electricity and high temperatures (Qu et al., 2022). In addition to preserving food, convective drying also helps in maintaining its flavor, aroma, and functional compounds. Convective drying, as a method of food preservation, combines innovation and sustainability by utilizing energy-efficient techniques to meet present food needs without compromising the ability of future generations to meet their own needs (Martynenko & Vieira, 2023). By utilizing convective drying, food can be efficiently preserved and dehydrated, reducing the risk of spoilage and ensuring its availability for consumption, contributing to food security and sustainability (Kaimal et al., 2022).

Overall, the integration of energy-efficient technologies like convective drying with sustainable practices is essential for maintaining a secure and environmentally conscious global food system that meets current needs without compromising the ability of future generation to do the same. This book chapter delves into the principles of drying, focusing on energy-efficient methods and their importance in the food industry. It explores various advanced drying technologies as well as the challenges faced in implementing these techniques. The chapter also discusses various strategies and approaches to overcome these challenges and improve the efficiency and sustainability of drying processes.

9.2 Energy-Efficient Approach to Food Dehydration

9.2.1 Exhaust Recirculation

Recirculating dehydrator exhaust air can lead to significant energy savings (Golman & Julklang, 2014). By mixing exhaust air with incoming fresh air, the temperature of the mixture is higher, requiring less energy to heat the mixture to the dryer's set point (Mujumdar & Menon, 2020). Studies have shown that recirculation rates of up to 75% can result in energy savings ranging from 45 to 50%, and recirculation rates from 60 to 85% can yield energy savings from 30 to 60%. In practical exhaust recirculation applications, airflow rates are adjusted by means of flow control valves, which provide more linear responses to airflow (Yu et al., 2022). Examples of these valves are knife-style dampers and louvre dampers. The dehydrator's internal measurements of humidity, wet bulb temperature, and dry bulb temperature can be used to inform control tactics (Eltawil et al., 2024). By installing ducting and valves to allow exhaust gas recirculation, retrofitting dehydrators for this feature can increase energy savings and overall operating efficiency.

9.2.2 Heat Pump Dehydration

Heat pump dehydrators, such as those offered by Nyle Dehydrators, precisely control temperature, humidity, and dwell time to reduce product moisture content while maintaining quality (Yousefi et al., 2023). These systems are highly efficient, using only the energy necessary to operate the compressor, blower, and circulating fans, resulting in up to 60% less energy than conventional dryers (Chojnacka et al., 2021). Solar Dehydration: Community solar dehydrators use a downdraft style with a solar collector on the south side of the dehydrator to warm air, which then enters the dehydration chamber and cools as it collects moisture from the food on the trays, sinking to the bottom of the chamber to be sucked out (Bakker & Slack, 2023). This method can be adapted to different needs and is suitable for large-scale passive dehydration without electricity. Osmotic Dehydration: Osmotic dehydration is a method for extracting juice from osmotically concentrated fruits without the need for heat treatment (Kaur et al., 2022). This process preserves the nutritional and organoleptic properties of the juice, making it a valuable technique for maintaining the quality of fruit products.

9.2.3 Infrared and Microwave Drying Techniques

Innovative drying methods like infrared, microwave and assisted drying methods offer energy-efficient solutions that preserve the chemical, taste, color, appearance and flavor components of dried products (Radojčin et al., 2021).

Infrared drying: Infrared drying utilities electromagnetic radiation to transfer heat directly to the surface of the material being dried (Bhagya et al., 2024). This method is efficient in removing moisture while preserving the products quality attributes. Infrared drying is particularly effective in applications where rapid and uniform drying is required, making it a valuable technique in food processing and other industries (Sakare et al., 2020).

Microwave drying: Microwave drying involves the use of electromagnetic waves to generate heat within the material being dried (Khodifad & Dhamsaniya, 2020). This method offers rapid and uniform heating, leading to shorter processing times and reduced energy consumption. Microwave drying is known for its ability to preserve the nutritional content, color, and flavor of food products, making it a popular choice in the food industry (Pandiselvam et al., 2022).

9.3 Convective Drying Principle

The principle of convective drying involves the use of heated air to remove moisture from various materials. In the context of food processing, convective drying is a crucial technique for preserving foodstuffs (Jayaraman & Gupta, 2020). This method is commonly applied in the drying of fruits, vegetables, and purees using spray dryers, where heated air interacts with the product's surface to facilitate moisture removal (Fathi et al., 2022). While convective drying offers advantages such as preventing overheating of products and preserving nutrients when cooled promptly after drying, it also presents challenges like oxidation of products and rapid loss of aromatic substances when in contact with air (Masud et al., 2020). In order to effectively transmit heat, convective drying in the food sector relies on direct contact between the dried item and warm air (Delfiya et al., 2022). Using heated air or gases, this technique removes moisture from food products, resulting in dehydrated goods that have a longer shelf life and a smaller volume (Pateiro et al., 2022). In order to effectively dry and preserve food goods, convective drying systems need a heat transfer mechanism. Warm air is blown around the food item during convective drying, absorbing moisture and removing it, which dehydrates the item. Convective drying is an effective method of removing moisture from food by employing heated air or gases (Solchansanj & Jayas, 2020). This method also uses less energy and preserves the food's flavor, aroma, and useful ingredients. This technique improves food security and sustainability in the sector by extending the shelf life of food and reducing dependency on non-renewable energy sources (Kaimal et al., 2022). The

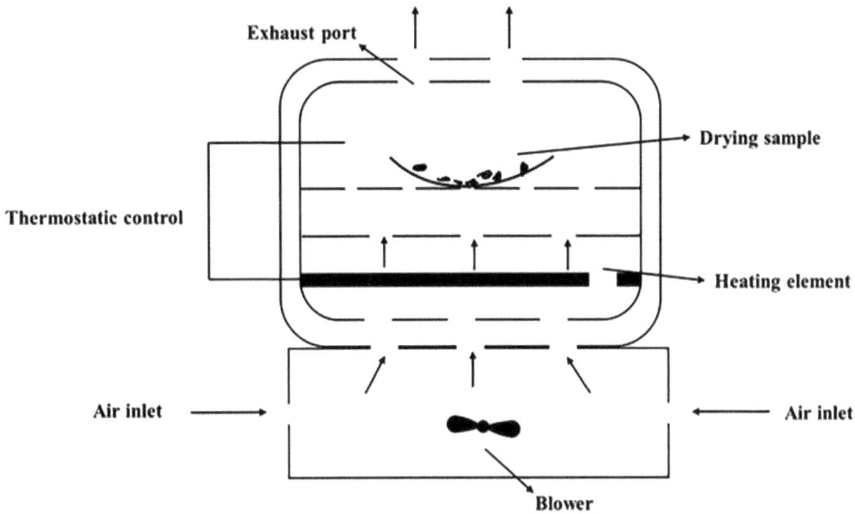

Fig. 9.1 Schematic diagram of natural convective drying used in the drying process

fundamental idea of convective drying in the food business is depicted in the diagram (Fig. 9.1).

9.4 Factors Affecting Efficiency of Convective Drying Process

Air Velocity: Air velocity, or the speed at which air moves over the material being dried, plays a crucial role in the efficiency of convective drying processes. Higher air velocity can enhance the rate of moisture removal, leading to faster drying times (Delfiya et al., 2022). This is because the increased air velocity promotes the evaporation of moisture from the material, which is then carried away by the air.

Drying Air Temperature: The temperature of the drying air is a critical factor influencing the efficiency of convective drying. Higher temperatures can accelerate the evaporation of moisture from the material, resulting in quicker drying (Babu et al., 2018).

Size and Shape of the Material: The size and shape of the material being dried impact the efficiency of convective drying. Smaller particles or thinner materials tend to dry more quickly compared to larger or thicker ones due to increased surface area exposure to the drying air (Mujumdar, 2002).

Moisture Content: The initial moisture content of the material affects the efficiency of convective drying. Materials with higher moisture content may require longer drying times and more energy to remove excess moisture (Kumar & Karim, 2019).

Heat Transfer Mechanisms: Understanding the heat transfer mechanisms involved in convective drying is essential for optimizing efficiency. Heat is transferred from the surrounding drying medium to the product through conduction and convection processes, influencing the overall drying rate (Wang et al., 2020).

Equilibrium between Product and Drying Medium: Achieving equilibrium between the product and the surrounding drying medium is crucial for efficient convective drying. Once this equilibrium is reached, mass and heat transfer cease, indicating the completion of the drying process (Kumar & Karim, 2019).

9.5 Impact of Convective Drying on Food Quality and Shelf Life

Convective drying, a common method used in the food industry for preserving food products, can have both positive and negative impacts on food quality and shelf life.

Positive Impacts of Convective Drying Include

- Improved Color: In some cases, convective drying can improve the color of food products, such as blackcurrant powder (Sadowska et al., 2019).
- Reduced Shrinkage: Compared to other drying methods, convective drying can result in less shrinkage, which can help maintain the product's structure (Calín-Sanchez et al., 2020).
- Preservation of Nutrients: Convective drying can help preserve nutrients, such as ascorbic acid, if the drying conditions are carefully controlled (Jayaraman & Gupta, 2020).

Negative Impacts of Convective Drying Include

- Loss of Volatiles: Convective drying can lead to a higher loss of volatile compounds compared to other drying methods (Calín-Sanchez et al., 2020).
- Reduction of Antioxidant Activity: Some food products may experience a significant reduction in antioxidant activity during convective drying (Zahoor & Khan, 2019).
- Formation of Off Flavors: Convective drying can generate off flavors, particularly in products like chokecherries, chokeberry, chokeberries, moringa leaves, and mango cubes (Calín-Sanchez et al., 2020).

To mitigate these negative impacts, it is essential to optimize the drying conditions, such as temperature, time, and slice thickness, to maintain the desired quality attributes of the food product (Deng et al., 2019). Additionally, the use of novel drying technologies, such as energy-saving combinations of existing techniques, can help improve the cost and quality of dried products while maintaining product quality (Menon et al., 2020).

9.6 Importance of Convective Drying on Food Security and Sustainability

Convective drying plays a crucial role in enhancing food security and sustainability through various mechanisms and impacts. Here are the key points highlighting the importance of convective drying to food security and sustainability:

Preservation and Shelf-Life Extension: Convective drying helps preserve foods beyond their natural shelf life by reducing water content and water activity, thus preventing degradation reactions of chemical, enzymatic, or microbial nature (ElGamal et al., 2023). This preservation method allows for diversifying the supply of foods with different flavors and textures, offering consumers a wide range of choices while buying foods.

Reduction of Energy Consumption: While drying is an energy-intensive process, advancements in convective drying technology have led to improved energy efficiency in food processing operations (Menon et al., 2020). By optimizing drying parameters and utilizing energy-efficient equipment, the energy consumption associated with convective drying can be reduced significantly (Chojnacka et al., 2021).

Minimization of Food Waste: Convective drying reduces waste by extending the shelf life of food products, thereby minimizing postharvest losses and contributing to food security. This preservation method adds value to food products, reduces transportation and storage costs, and increases the overall shelf life of dried foods.

Nutrient Retention: Despite some nutrient loss during drying processes, efforts are made to maximize nutrient retention during convective drying (Hasan et al., 2019). Strategies such as optimizing drying conditions and using advanced techniques aim to preserve macronutrients, micronutrients, and bioactive compounds in dried foods.

Enhanced Food Quality: Convective drying helps maintain the quality of dried foods by minimizing chemical changes like enzymatic browning and maximizing nutrient retention during the process (Jayaraman & Gupta, 2020). By focusing on quality aspects such as texture, color, flavor, and nutritional content, convective drying contributes to sustainable food production practices.

9.7 Enhancing Sustainable Food Practices with Advanced Drying Technology

Enhancing sustainable food practices with advanced drying technology is crucial for achieving global food security and sustainability. Advanced drying technology plays a significant role in preserving food products, reducing waste, and improving energy efficiency. Here are some key points highlighting the importance of advanced drying technology in sustainable food practices:

Energy Efficiency: Advanced drying technologies focus on improving energy efficiency in the drying process, reducing energy consumption, and minimizing environmental impact (Lamidi et al., 2019). By utilizing innovative techniques like heat pumps, superheated steam drying, and vacuum drying, energy consumption can be optimized, leading to more sustainable food processing operations (Acar et al., 2022).

Quality Preservation: Advanced drying technology aims to minimize quality deterioration during the drying process by optimizing parameters such as temperature, air velocity, and humidity control (Fathi et al., 2022). By preserving the texture, color, flavor, and nutritional content of dried foods, these technologies contribute to maintaining high product quality and consumer satisfaction (Zhang et al., 2022).

Reduction of Food Waste: By extending the shelf life of food products through advanced drying methods, the need for refrigeration systems is reduced, leading to lower transportation and storage costs (Sridhar et al., 2021). This reduction in food waste contributes to sustainable food practices by minimizing postharvest losses and ensuring a more efficient food supply chain.

Innovative Techniques: Emerging dehydration technologies like microwave drying, infrared drying, and osmotic dehydration offer more sustainable processing methods by reducing drying time and improving energy efficiency. These innovative techniques contribute to a more environmentally friendly approach to food dehydration while maintaining product quality (Pandiselvam et al., 2022).

Global Food Security: Advanced drying technology plays a vital role in enhancing global food security by preserving perishable foods, reducing losses during storage and transportation, and ensuring a stable food supply chain (Onwude et al., 2020). By adopting sustainable drying practices, countries can improve their food security status and reduce dependency on imports.

By embracing innovative drying techniques and technologies, the food industry can move towards a more sustainable and environmentally conscious approach to food processing and preservation.

9.8 Challenges and Opportunities in Food Drying Processes

Food drying processes face various challenges that impact product quality, energy efficiency, and overall sustainability. However, these challenges also present opportunities for innovation and improvement in the food industry. Here are some key challenges and opportunities in food drying processes.

Challenges

Quality Deterioration: Drying can lead to quality deterioration in food products, affecting attributes like color, flavor, and texture (Verma et al., 2020).

Energy Intensity: Drying is an energy-intensive process, contributing to high operational costs and environmental impact (Qu et al., 2022).

Microbial Growth: Inadequate drying conditions can lead to microbial growth and spoilage of food products (Fathi et al., 2022).

Nutrient Loss: Some drying methods may result in the loss of essential nutrients in food products (Hasan et al., 2019).

Opportunities

Optimization of Drying Conditions: By optimizing drying parameters such as temperature, airflow, and humidity, it is possible to enhance product quality and reduce energy consumption (Majdi et al., 2019).

Incorporation of Pre-treatments: Pre-treatments before drying can improve the quality and shelf life of food products by enhancing moisture removal efficiency (Santos et al., 2021).

Integration of Nanotechnology: Nanotechnology applications in drying processes can offer innovative solutions for improving product quality and reducing energy consumption (Adeyeye et al., 2022).

Artificial Intelligence (AI) Integration: Implementing AI for real-time quality monitoring systems can help optimize drying processes, enhance product quality, and reduce waste (Iheonye et al., 2023).

Development of Hybrid Drying Methods: Combining different drying techniques can lead to more efficient and sustainable drying processes with improved product quality (Qu et al., 2022).

Enhanced Quality Prediction Models: Advanced modeling techniques can help predict and control food quality during the drying process, ensuring consistent product characteristics (Ajuebor et al., 2020).

9.9 Strategies for Implementing Convective Drying in Food Industries

Implementing convective drying in food industries requires a strategic approach to optimize the process for efficiency, quality, and sustainability. Here are key strategies based on the provided sources:

Optimizing Drying Parameters: Fine-tuning parameters like temperature, air velocity, and humidity control are essential for efficient convective drying processes (Depiver & Mallik, 2023). This optimization helps enhance product quality and reduce energy consumption.

Incorporating Pre-Treatments: Pre-treatments before drying can improve the efficiency of convective drying by enhancing moisture removal and preserving the quality of food products (Liu et al., 2021).

Utilizing Advanced Equipment: Employing specialized equipment like spray dryers for convective drying can offer automation, scalability, and precise control over the drying process (Samborska et al., 2022). This equipment can be adjusted to different work volumes and operate in an automatic mode, saving on maintenance costs.

Addressing Disadvantages: Despite its advantages, convective drying has drawbacks such as oxidation of products and loss of aromatic substances (Jayaraman & Gupta, 2020). Strategies to mitigate these disadvantages include optimizing air temperature, cooling products promptly after drying, and considering alternative drying methods for sensitive products.

Energy Efficiency Measures: Implementing energy-efficient practices like heat recovery systems and optimizing air circulation can reduce energy consumption during convective drying processes (Majdi et al., 2019).

Sustainability Assessment: Evaluating the sustainability of convective drying processes through indices like the Sustainability Index (SI) and Renewability Index (RI) can guide improvements in energy efficiency, environmental impact, and overall sustainability of food drying technologies (Martynenko et al., 2023).

9.10 Future of Food Security

Convective drying in the food industry is a critical process that involves removing water from food substances using heat. This method is employed to preserve food products, reduce moisture content, and extend shelf life. Recent studies have shown that convective drying can deliver better product quality, especially at high temperatures, although challenges like scorching of food surfaces and poor color retention may arise. Advancements in convective drying techniques, such as far-infrared drying, are expected to play a significant role in the future of food security. Far-infrared drying offers benefits like low energy consumption and improved quality retention, making it a promising method for sustainable food processing. Overall, the evolution of drying technologies, including convective drying, is crucial for addressing global concerns related to food security and sustainability. By continuously improving drying processes and adopting innovative techniques, the food industry can enhance efficiency, reduce waste, and contribute to a more sustainable and secure food supply chain.

9.11 Conclusion

In conclusion, convective drying is a crucial method for preserving food products, extending their shelf life, and reducing food waste. By utilizing energy-efficient techniques like convective drying, we can maintain food security and sustainability while minimizing our reliance on non-renewable energy sources. This chapter has explored the principles of drying, focusing on energy-efficient methods and their importance in the food industry. It has also discussed various advanced drying technologies and the challenges faced in implementing these techniques. By understanding and addressing these challenges, we can improve the efficiency and sustainability of drying processes and contribute to a more secure and environmentally conscious global food system.

References

Acar, C., Dincer, I., & Mujumdar, A. (2022). A comprehensive review of recent advances in renewable-based drying technologies for a sustainable future. *Drying Technology, 40*(6), 1029–1050.

Adeyeye, S. A. O., Ashaolu, T. J., & Babu, A. S. (2022). Food drying: A review. *Agricultural Reviews, 1*(8).

Ajuebor, F., Aworanti, O., Agbede, O., Agarry, S., Afolabi, T., & Ogunleye, O. (2020). Convective hot air drying of chilli pepper: Process optimization and modelling the drying kinetics and quality attributes of dried product.

Babu, A. K., Kumaresan, G., Raj, V. A. A., & Velraj, R. (2018). Review of leaf drying: Mechanism and influencing parameters, drying methods, nutrient preservation, and mathematical models. *Renewable and Sustainable Energy Reviews, 90*, 536–556.

Bakker, E., & Slack, G. (2023). *An island called California: an ecological introduction to its natural communities*. Univ of California press.

Bhagya, J., Reshmi, S. K., Waghmare, R., Moses, J. A., & Anandharamakrishnan, C. (2024). Electromagnetics and its allied applications in food processing. In *Emerging technologies for the food industry* (pp. 185–234). Apple Academic Press.

Calín-Sánchez, Á., Lipan, L., Cano-Lamadrid, M., Kharaghani, A., Masztalerz, K., Carbonell-Barrachina, Á. A., & Figiel, A. (2020). Comparison of traditional and novel drying techniques and its effect on quality of fruits, vegetables and aromatic herbs. *Foods, 9*(9), 1261.

Chojnacka, K., Mikula, K., Izydorczyk, G., Skrzypczak, D., Witek-Krowiak, A., Moustakas, K., Kułażyński, M., et al. (2021). Improvements in drying technologies-Efficient solutions for

cleaner production with higher energy efficiency and reduced emission. *Journal of Cleaner Production, 320*, Article 128706.
Clapp, J., Moseley, W. G., Burlingame, B., & Termine, P. (2022). The case for a six-dimensional food security framework. *Food Policy, 106*, Article 102164.
Delfiya, D. A., Prashob, K., Murali, S., Alfiya, P. V., Samuel, M. P., & Pandiselvam, R. (2022). Drying kinetics of food materials in infrared radiation drying: A review. *Journal of Food Process Engineering, 45*(6), Article e13810.
Deng, L. Z., Mujumdar, A. S., Zhang, Q., Yang, X. H., Wang, J., Zheng, Z. A., Xiao, H. W., et al. (2019). Chemical and physical pretreatments of fruits and vegetables: Effects on drying characteristics and quality attributes–a comprehensive review. *Critical Reviews in Food Science and Nutrition, 59*(9), 1408–1432.
Depiver, J. A., & Mallik, S. (2023). An empirical study on convective drying of ginger rhizomes leveraging environmental stress chambers and linear heat conduction methodology. *Agriculture, 13*(7), 1322.
ElGamal, R., Song, C., Rayan, A. M., Liu, C., Al-Rejaie, S., & ElMasry, G. (2023). Thermal degradation of bioactive compounds during drying process of horticultural and agronomic products: A comprehensive overview. *Agronomy, 13*(6), 1580.
Eltawil, M. A., Azam, M. M., & Mohammed, M. (2024). Applications of Solar Energy for Enhancing Sustainable Food. In *Food and nutrition security in the Kingdom of Saudi Arabia, Vol. 1: National analysis of agricultural and food security* (pp. 263–318). Springer International Publishing.
Fathi, F., Ebrahimi, S. N., Matos, L. C., Oliveira, P. P., M. B., & Alves, R. C. (2022). Emerging drying techniques for food safety and quality: A review. *Comprehensive Reviews in Food Science and Food Safety, 21*(2), 1125–1160.
García-Díez, J., Gonçalves, C., Grispoldi, L., Cenci-Goga, B., & Saraiva, C. (2021). Determining food stability to achieve food security. *Sustainability, 13*(13), 7222.
Golman, B., & Julklang, W. (2014). Analysis of heat recovery from a spray dryer by recirculation of exhaust air. *Energy Conversion and Management, 88*, 641–649.
Hasan, M. U., Malik, A. U., Ali, S., Imtiaz, A., Munir, A., Amjad, W., & Anwar, R. (2019). Modern drying techniques in fruits and vegetables to overcome postharvest losses: A review. *Journal of Food Processing and Preservation, 43*(12), Article e14280.
Iheonye, A. C., Raghavan, V., Ferrie, F. P., Orsat, V., & Gariepy, Y. (2023). Monitoring visual properties of food in real time during food drying. *Food Engineering Reviews, 15*(2), 242–260.
Jayaraman, K. S., & Gupta, D. D. (2020). Drying of fruits and vegetables. In *Handbook of industrial drying* (pp. 643–690). CRC Press.
K., Izydorczyk, G., Skrzypczak, D., Witek-Krowiak, A., Moustakas, K., & Kułażyński, M. et al. (2021). Improvements in drying technologies-Efficient solutions for cleaner production with higher energy efficiency and reduced emission. *Journal of Cleaner Production, 320*, 128706.
Kaimal, A. M., Tidke, V. B., Mujumdar, A. S., & Thorat, B. N. (2022). Food security and sustainability through solar drying technologies: A case study based on solar conduction dryer. *Materials Circular Economy, 4*, 1–23.
Kaur, D., Singh, M., Zalpouri, R., & Singh, I. (2022). Osmotic dehydration of fruits using unconventional natural sweeteners and non-thermal-assisted technologies: A review. *Journal of Food Processing and Preservation, 46*(12), Article e16890.
Khodifad, B. C., & Dhamsaniya, N. K. (2020). Drying of food materials by microwave energy-A review. *International Journal of Current Microbiology and Applied Sciences, 9*(5), 1950–1973.
Kumar, C., & Karim, M. A. (2019). Microwave-convective drying of food materials: A critical review. *Critical Reviews in Food Science and Nutrition, 59*(3), 379–394.
Lamidi, R. O., Jiang, L., Pathare, P. B., Wang, Y., & Roskilly, A. P. (2019). Recent advances in sustainable drying of agricultural produce: A review. *Applied Energy, 233*, 367–385.
Liu, Z. L., Zielinska, M., Yang, X. H., Yu, X. L., Chen, C., Wang, H., Xiao, H. W., et al. (2021). Moisturizing strategy for enhanced convective drying of mushroom slices. *Renewable Energy, 172*, 728–739.

Majdi, H., Esfahani, J. A., & Mohebbi, M. (2019). Optimization of convective drying by response surface methodology. *Computers and Electronics in Agriculture, 156*, 574–584.

Martynenko, A., & Vieira, G. N. A. (2023). Sustainability of drying technologies: System analysis. *Sustainable Food Technology, 1*(5), 629–640.

Masud, M. H., Karim, A., Ananno, A. A., & Ahmed, A. (2020). *Sustainable food drying techniques in developing countries: Prospects and challenges* (p. 20220004385). Springer.

Menon, A., Stojceska, V., & Tassou, S. A. (2020). A systematic review on the recent advances of the energy efficiency improvements in non-conventional food drying technologies. *Trends in Food Science & Technology, 100*, 67–76.

Mujumdar, A. S. (2002). Drying research-current state and future trends. *Developments in Chemical Engineering and Mineral Processing, 10*(3–4), 225–246.

Mujumdar, A. S., & Menon, A. S. (2020). Drying of solids: principles, classification, and selection of dryers. In *Handbook of industrial drying* (pp. 1–39). CRC Press.

Onwude, D. I., Chen, G., Eke-Emezie, N., Kabutey, A., Khaled, A. Y., & Sturm, B. (2020). Recent advances in reducing food losses in the supply chain of fresh agricultural produce. *Processes, 8*(11), 1431.

Pandiselvam, R., Tak, Y., Olum, E., Sujayasree, O. J., Tekgül, Y., Çalışkan Koç, G., Kumar, M., et al. (2022). Advanced osmotic dehydration techniques combined with emerging drying methods for sustainable food production: Impact on bioactive components, texture, color, and sensory properties of food. *Journal of Texture Studies, 53*(6), 737–762.

Pateiro, M., Vargas-Ramella, M., Franco, D., da Cruz, A. G., Zengin, G., Kumar, M., Lorenzo, J. M., et al. (2022). The role of emerging technologies in the dehydration of berries: Quality, bioactive compounds, and shelf life. *Food Chemistry: X, 16*, Article 100465.

Qu, H., Masud, M. H., Islam, M., Khan, M. I. H., Ananno, A. A., & Karim, A. (2022). Sustainable food drying technologies based on renewable energy sources. *Critical Reviews in Food Science and Nutrition, 62*(25), 6872–6886.

Radojčin, M., Pavkov, I., Bursać Kovačević, D., Putnik, P., Wiktor, A., Stamenković, Z., Gere, A., et al. (2021). Effect of selected drying methods and emerging drying intensification technologies on the quality of dried fruit: A review. *Processes, 9*(1), 132.

Sadowska, A., Rakowska, R., Świderski, F., Kulik, K., & Hallmann, E. (2019). Properties and microstructure of blackcurrant powders prepared using a new method of fluidized-bed jet milling and drying versus other drying methods. *CyTA-Journal of Food, 17*(1), 439–446.

Sahoo, M., Titikshya, S., Aradwad, P., Kumar, V., & Naik, S. N. (2022). Blanching, pasteurization and sterilization: principles and applications. *Thermal Food Engineering Operations*, 75–115.

Sakare, P., Prasad, N., Thombare, N., Singh, R., & Sharma, S. C. (2020). Infrared drying of food materials: Recent advances. *Food Engineering Reviews, 12*(3), 381–398.

Samborska, K., Poozesh, S., Barańska, A., Sobulska, M., Jedlińska, A., Arpagaus, C., Jafari, S. M., et al. (2022). Innovations in spray drying process for food and pharma industries. *Journal of Food Engineering, 321*, Article 110960.

Santos, K. C., Guedes, J. S., Rojas, M. L., Carvalho, G. R., & Augusto, P. E. D. (2021). Enhancing carrot convective drying by combining ethanol and ultrasound as pre-treatments: Effect on product structure, quality, energy consumption, drying and rehydration kinetics. *Ultrasonics Sonochemistry, 70*, Article 105304.

Solchansanj, S., & Jayas, D. S. (2020). Drying of foodstuffs. In *Handbook of industrial drying* (pp. 589–625). CRC Press.

Sridhar, A., Ponnuchamy, M., Kumar, P. S., & Kapoor, A. (2021). Food preservation techniques and nanotechnology for increased shelf life of fruits, vegetables, beverages and spices: A review. *Environmental Chemistry Letters, 19*, 1715–1735.

Tan, X., Zhu, X., & Tang, J. (2023). Postharvest strategies—efficiency and novel processes for extending shelf life and reducing spoilage potential for minimizing food waste. In *Future proteins* (pp. 223–240). Academic Press.

Verma, D. K., Thakur, M., Srivastav, P. P., Karizaki, V. M., & Suleria, H. A. R. (2020). Effects of drying technology on physiochemical and nutritional quality of fruits and vegetables.

In *Emerging thermal and nonthermal technologies in food processing* (pp. 69–116). Apple Academic Press.
Wang, G., Wu, W., Qiao, F., Fu, D., Liu, Z., & Han, F. (2020). Research on an electric energy-saving grain drying system with internal circulation of the drying medium. *Journal of Food Process Engineering, 43*(9), Article e13476.
Wang, Y., Yuan, Z., & Tang, Y. (2021). Enhancing food security and environmental sustainability: A critical review of food loss and waste management. *Resources, Environment and Sustainability, 4*, Article 100023.
Yousefi, S., Weisany, W., Hosseini, S. E., & Ghasemlou, M. (2023). Mechanisms of nanoencapsulation to boost the antimicrobial efficacy of essential oils: A review. *Food Hydrocolloids*, 109655.
Yu, Z. H. O. U., Xueyu, L. I., Shuiting, D. I. N. G., Shuai, Z. H. A. O., Kun, Z. H. U., Longtao, S. H. A. O., & Zheng, X. U. et al. (2022). Technologies and studies of gas exchange in two-stroke aviation piston engine: A review. *Chinese Journal of Aeronautics.*
Zahoor, I., & Khan, M. A. (2019). Microwave assisted convective drying of bitter gourd: Drying kinetics and effect on ascorbic acid, total phenolics and antioxidant activity. *Journal of Food Measurement and Characterization, 13*, 2481–2490.
Zhang, Y., Pandiselvam, R., Zhu, H., Su, D., Wang, H., Ai, Z., Liu, Y., et al. (2022). Impact of radio frequency treatment on textural properties of food products: An updated review. *Trends in Food Science & Technology, 124*, 154–166.

Part IV
Overarching Themes

Chapter 10
Adaptation to Climate Change as a Pre-requisite to Food and Water Security

Nagma Khan, Baby Tabassum, Asma Hasan, and Mohammad Hashim

Abstract Climate change adaptation refers to activities taken to mitigate susceptibility to existing or predicted impacts from climate change, including weather extremes and hazards, elevated sea levels, diminished biodiversity, and food and water shortages. Food and water security are the most pressing issues under climate change due to their high susceptibility to constantly changing climatic conditions. Change in the climate severely impacts on agricultural lands, increasing or decreasing food security and, consequently, food nutrition. This will have far-reaching consequences for agricultural water management, even if adaptation capability is relatively good. This is because the more hunger insecurity there is, the more instances of malnutrition are recorded. In developing nations, the effects will vary greatly by region, but they will be caused by a mixture of less beneficial plant development circumstances such as more erratic rainfall, reduced accessibility of water for farming, and increased crop water consumption. These stressors will be added to the pressures to increase the production of food while using less water and reducing deterioration of soil in the face of expanding global populations and shifting dietary habits. The paper emphasizes the management of water-related concerns, agricultural security, and the impact of climate change and provides a more coherent structure.

Keywords Biodiversity loss · Water shortages · Adaptation capability · Nature-based solutions

---

N. Khan · B. Tabassum (✉) · A. Hasan · M. Hashim
Department of Zoology, Toxicology Lab, Govt. Raza P.G. College, Rampur, UP, India
e-mail: dr.btabassum@gmail.com

N. Khan · B. Tabassum · A. Hasan
Mahatma Jyotiba Phule Rohilkhand University, Bareilly, UP, India

J. Parray and A. K. Haghi (eds.), *Food and Water Security*,
https://doi.org/10.1007/978-3-032-04870-7_10

10.1 Introduction

Climate variability transpires when alterations in the world's climatic cycle produce novel weather phenomena that endure for at least many decades, perhaps extending for billions of years. The effects of changing the climate are evident in the increased incidence and extent of severe weather situations, such as flooding, drought, altered precipitation events, and extreme climatic conditions. Rising temperatures have resulted in a proliferation of weeds, pests, and illnesses, posing a hazard to marine life, flora, fauna, and humans. Similarly, a rise in extreme weather patterns have contributed in erosion of soil due to winds and water, as well as a reduction in fertile land for forestry and agriculture. The worldwide food system, which sustains over 1 billion individuals, is essential for the majority of the populace and contributes between 1 and 60% to national GDP. Since 1961, per capita food supply has risen by more than 30%, mostly attributable to enhanced utilization of nitrogen fertilizers and water resources. Climate change directly affects food systems and food security, potentially intensifying competition for agricultural resources. To mitigate climate change, land-based negative emissions technologies, such as biomass production, may exert pressure on agricultural output and security of food as a result of possible land competition. The effect of climate change on water supplies and agriculture necessitates the utilization of simulation models to forecast the location and magnitude of alterations in vital factors that regulate growth of crops (temperature and evaporative demand) and water availability (precipitation, evaporation, streamflow, and groundwater recharge). Agricultural water management includes all technology and techniques that maintain optimal soil moisture for plant development, ranging from improving rainwater capture and retention to comprehensive irrigation in the absence of rainfall. It encompasses drainage supplies and the mitigation and avoidance of flooding. The expected effects of changes in the climate exert more strain on food production systems required to satisfy the needs of a swiftly expanding and increasingly prosperous worldwide population. As agriculture advances and intensifies the utilization of resources like water and soil, its effects on ecological systems become increasingly apparent. Impairing the sustainability of these habitats jeopardizes the culinary cultivation processes they sustain. The evaluation of feasible and efficient solutions to the effects of environmental change on rainfall and agribusiness necessitates extensive knowledge and integration of agronomic research, management of water, and hydrology. Consideration of the resultant environmental interactions and trade-offs will be crucial (Turral et al., 2011).

Climate change resilience emphasizes water as an essential element in mitigating adverse weather occurrences. There are three main groups of climate variable impacts: insufficient water (droughts), excessive amount of water (floods), and highly contaminated water. Cities worldwide are currently experiencing these effects in water-intensive sectors, including agriculture, hydropower, fisheries, and industry. The 2030 Global Agenda includes specific objectives for water and climate change, underscoring the interconnection between the two. Improving water security is necessary to address water-related issues that are a result of variations in the climate. The notion of "water security" refers to the ability of individuals to ensure feasible utilization of sufficient amounts of water of suitable quality in order to sustain existence, wellness, and economic growth, as well as to protect ecosystems from water-borne contaminants and disasters related to water to maintain a climate of harmony and prosperity. Governments and policymakers worldwide prioritize the establishment of a water-secure society. In recent decades, water security has garnered a greater amount of attention and is a critical component of contemporary science and policy agendas (Babel et al., 2020). The United Nations Development Program cautions that changes in climate may impede or even reverse the advancements people have made throughout the previous decade concerning the accessibility of food in water, crop yield and availability, nutrition, and healthcare. By 2080, the consequences of changing climate—such as extreme heat, drought, and increasing sea levels, are fluctuations in rainfall—may result in an additional 600 million individuals experiencing hunger and elevate the population confronting water shortages by 1.8 billion (Opschoor, 2008). The effect of environmental change on the purity of water is significant, underscoring the critical necessity for robust adaptation strategies. Variations in precipitation, rising temperatures, and extreme weather occurrences significantly contribute to the decline in water quality. Intense rainfall can lead to increased drainage of storm water, presenting challenges to water quality and complicating treatment processes. Dry spells can result in the accumulation of pollutants, exacerbating the decline in quality of water (US EPA, 2024a). In coastal areas, increasing sea levels may result in the encroachment of saltwater. The emergence of harmful algal blooms can exacerbate these issues. In light of these challenges, various strategies are currently being implemented, including advancements in water treatment technologies, the adoption of economically viable watershed governance methods, continuous monitoring of water quality, and the rehabilitation of ecological systems like wetland systems. A thorough approach that combines innovation in technology, effective management practices, and engaged community participation is crucial for safeguarding public health and ensuring sustainable water resources.

Food security is "a condition in which all individuals, at all times, have physical, social, and economic access to a sufficient, safe, and nutritious diet that satisfies dietary requirements and food preferences for an active and healthy life." Food safety is not just characterized by nutrition availability; rather, it is the extent to which the financial as well as non-economic sources available to the individuals are sufficient to ensure that all individuals have the ability to sufficient amount and quality of food. It is important to ensure that food safety will be contingent upon financial growth, modifications to global trade, inventories, and assistance for food aid policies, in addition to the socioeconomic and environment-related influences on agricultural output.

10.2 The Effects of Climate Change on Food Safety

The agricultural economy of India is facing important issues resulting from changes in the climate, which poses a growing hazard to food safety. Agricultural output in India is highly susceptible to climatic vicissitudes, which include irregular temperatures, aberrant precipitation levels, and humidity fluctuations. The evidence indicates that agricultural output and security of food are significantly affected by the rising number and intensity of severe weather incidents (e.g., water scarcity, heatwaves). According to research, the productivity of wheat decreases by 4–5 million metric tons for each 1 °C rising temperature. A plethora of data exists on how warmer temperature impacts the yield and nutritional value of numerous agricultural products, including food crops, therapeutic plants, dairy, fish, and beverages (such as coffee and tea). The farming sector along with production, transportation, storage, and accessibility to food is affected by the current challenges to Indian food security. To attain sustainable food security, India's primary objective is to enhance food production levels and guarantee equitable access across the country, particularly within the context of climate crisis. Food safety is a complex and multifaceted subject that encompasses quality, purchasing power, sustainability, distribution, and food synthesis. The socioeconomic repercussions of climate variations include diminished earnings, compromised livelihoods, trade disruptions, and negative health outcomes, as it impacts food safety through its impact on crops, cattle, forests, fisheries, and aquaculture. The magnitude of the climatic disturbance and its inherent vulnerabilities determine the overall impact of climate change. From 1981 to 1991, India's population growth was 24%, but it reduced to 21.1% from 1991 to 2001. The growth rate of dietary grains, which encompasses barley, bran, legumes, and other grains, decreased from 2.8 to 0.94% during this time (Dev & Sharma, 2010). This decline is concerning and requires substantial enhancements in food and cereal production to correspond with population growth. The agricultural sector in India experienced a decline in growth from 3.5% in 1981–1982 to 1996–1997 to approximately 2% in 1997–1998 to 2004–2005 (Chand et al., 2014). The agricultural sector and livelihoods are adversely affected by the lethargic growth, which poses ample threat to national food safety. The agricultural industry experienced a 3.75% development rate from 2004–2005 to 2012–2013 as a consequence of the implementation of measures to revitalize it. In order to satisfy the growing population, grain output must increase by 50% by 2020. Due to rapid population increase and industrialization, per capita agricultural land has declined from 0.49 ha in 1950 to 0.16 ha in 2000. This number is expected to drop to 0.09 ha by 2020. Indian farming and food manufacturing are particularly susceptible to changes in the climate, primarily as a result of their susceptibility to monsoon fluctuations, with wheat and rice being particularly affected. However, the decrease in cultivation of thermosensitive commodities, such as tea, cardamom, coffee, black pepper, cashew, and cocoa, is being caused by terminal heat damage and moisture in the soil storage. Mango and cashew flowering will be influenced by night temperatures, whereas coconut output may be impacted by heat waves and flooding. Plants are the predominant source of protein in India due to destitution

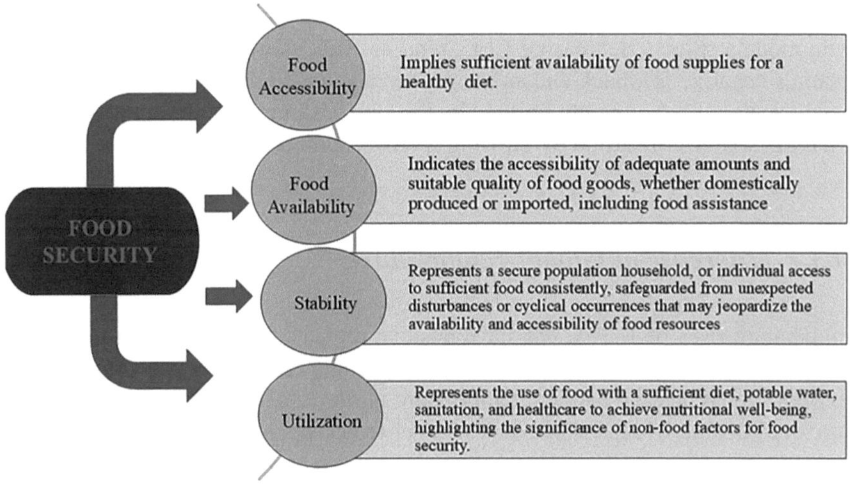

Fig. 10.1 The UN FAO defines the four aspects of food safety

and cultural considerations. This could potentially result in the average Indian losing over 6 percent of their everyday protein and exposing 53 million individuals to additional risks as the amino acid content in rice, wheat, barley, and potato decreases as atmospheric carbon dioxide increases (Fig. 10.1) (Medek et al., 2017).

10.3 Effects of CO_2 on Plant Fertilization

Increased CO_2 levels are boosting harvest yields as a result of change in climate impacts on plant development and growth in temperate zones. For crops that are very efficient at photosynthetic processes, this impact can reach 0–10%; for crops that are less efficient, it is larger. Based on current projections, atmospheric CO_2 levels are expected to reach 550 ppm (Ajilogba & Walker, 2021). The CO_2 fertilization effect (CFE), stemming from elevated atmospheric carbon dioxide levels, enhances photosynthesis in plants while constraining leaf transpiration. This impact is contingent upon plant type, ambience, and soil temperature, as well as the availability of nutrients in water (Ueyama et al., 2020). Net primary production may favorably respond to the impact; nevertheless, it does not directly augment plant growth or carbon sequestration. The influence has contributed to a 44% growth in GDP since the 2000s. Global resource designs, terrestrial component designs, and innovative worldwide flora systems are employed to analyze plant diversity in relation to CO_2 concentrations; nevertheless, the ecological processes linked to this effect remain ambiguous and difficult to predict. Although the CFE—the increased plant production caused by higher CO_2 concentrations—maintains significant negative feedback on global warming, its temporal dynamics are still not well understood.

Even though it is far less than the trend shown in global observations, current carbon cycle models show a decreasing CFE trend as well. A weakening of the climatic system's negative feedback and an increase in society's reliance on future methods to reduce global warming are implied by this diminishing increase in the regulation of natural sources of carbon by growing levels of airborne CO_2.

10.3.1 Increase of Global Temperature

Temperature rises of 1–3° C can have positive impacts on plants in warm regions, while tropical and occasionally arid areas, especially cereal-growing countries, see the opposite. When the average temperature is more than 3° C, the effect is even more detrimental. Because heat waves hinder feed crop development, which in turn impacts cattle output, rising temperatures have indirect consequences. Sunlight has direct consequences on the rate of heat transfer between living things and their natural surroundings. The daily existence of more than 180 million people who produce cattle on rangelands in Africa is at risk due to the sensitivity of forage production from these areas to climate change. Additionally, there are substantial climatic impacts on digestibility, fodder quality, nutritional content, and animal performance (Ajilogba & Walker, 2021).

During the Industrial era, anthropogenic activities, particularly the production of greenhouse gases, have been the main reason for the Earth's air temperature rising. The wide range of the heating has occurred after 1975, according to NASA, the worldwide average temperature has risen by 1.2 °C since 1880 (Hansen et al., 2010). Variations in local and short-term temperatures can occur as a result of both predictable and unpredictable phenomena, such as changes in wind and precipitation patterns. Both the quantity of energy that the Earth gets from the Sun and the amount of energy that it emits back into space determine the average global temperature. A planet's radiative output is proportional to its atmospheric chemical make-up, and more specifically to the concentration of CO2, which absorbs heat. The quantity of energy needed to warm the seas, environment, and surfaces of land equals a 1° global shift, which is a substantial amount (NASA Earth Observatory, 2022).

10.4 The Impacts of Changing Climate on Food Accessibility

The availability of food is the capacity to control and possess food, disseminate it, and make food choices, thereby allowing individuals to effectively convert their desire into a demand. The nature of society of food availability, hunger, and susceptibility are critical aspects of food accessibility, as they entail the ability of communities and individuals to spend money. This concept is crucial for guaranteeing that all individuals have access to sustenance and that it is distributed fairly (Sage, 2014). World's shortage of food crisis has been especially influenced by climate crisis, with models indicating that an additional 200–600 million individuals may experience famine by 2080. It is anticipated that changing climate will have a substantial impact on future hunger, even when economic growth is taken into account. By 2050, severe stunting is expected to improve by 62.2% in South Asia and by 54.5% in east and south sub-Saharan Africa. Changing climate has been a matter of discussion in India, with specialists such as Ramachandran (2014) cautioning that it has the potential to significantly impede or reverse the progress that has been made in the places of food safety and nutrition. The economic determinants of food insecurity are exacerbated by changing climate, which results in a higher frequency of extreme events and fluctuation in crop-producing seasons, which negatively impacts the net incomes of producers. Small and marginal farmers in rural India depend on rain-fed monocropping, which only provides a few months of food security. Climate change also harms fishermen and forest-dependent people, with landless farmers at greatest danger of losing food. In locations with substantial food insecurity and inequality, droughts and flooding will affect children more owing to their susceptibility. Food instability in urban areas is also a significant concern, as impoverished individuals from the coast and suburban areas frequently relocate to urban areas in search of stable livelihoods. Household purchasing power significantly influences this, whereas social dynamics related to food access, poverty, and susceptibility are essential factors. This frequent purchase, barter, or swap of food is influenced by climate, employment prospects, and the availability of certain food items. Climate can result in elevated pricing, rendering certain items costly and adversely affecting human nutrition and health. Low-income households frequently bear the brunt of these increases, resulting in diminished consumption or a decline in the intake of favorite meals. Consequently, food prices are essential for food security (Ajilogba & Walker, 2021). A surge of migration toward cities is frequently precipitated by hunger, which forces entire families to relocate to urban neighborhoods. There they become part of the informal urban sector, where they are paid inadequately (Lloyd et al., 2011). The urban food insecurity indicators of India are alarming, with over 30% of children under the age of five in urban Bihar, Madhya Pradesh, and Karnataka being underweight.

10.5 Water System and Security Effected by Climate Mitigation

Climate change encompasses a multifaceted mixture of gradual patterns, including temperature increases, alterations in variability, and shifts in severe occurrences. Their substantial environmental and socioeconomic consequences are expected to emerge through effects on the hydrological cycle and security of water. Alterations in precipitation timing, intensity, and form, together with atmospheric evaporative demand, will modify the accessibility of surface and groundwater resources. Land surface attributes, encompassing use of land and surface coverage, can influence alterations in the water cycle. The consequences of glacier and snow cover reduction in mountainous regions demonstrate intricate relationships between water supply systems and glacial meltwater. Factors such as climate change, growth in population, and industrialization will influence the requirement for water and future water security. Although there is considerable confidence in the continuation of climate change, there is diminished clarity about the evolution of critical components of the water cycle. Future precipitation patterns exhibit significant variability owing to discrepancies among climate model outcomes and the potential for decade-long fluctuations in precipitation. Nonetheless, there will be an elevated frequency and intensity of floods, along with an increase in worldwide precipitation, attributable to rising temperatures.

Water safety depends on their being enough for residential and non-domestic use. Additionally, it helps preserve and restore natural habitats. Some definitions highlight humans, while others recognize inter- and intra-generational equality with the natural environment. Aquifers, glaciers, and ice sheets must be used to store water for people in different nations and locations with different socioeconomic and cultural conditions (Water Security under Climate Change, 2021). Governments and legislators worldwide have prioritized water safety in recent years. Researchers and organizations have developed several water security definition and measurement methods to improve it. The scope of water security was first defined as safeguarding the food supply, satisfying essential necessities, conserving ecosystems, sharing water resources, managing risks, valuing water, and controlling water wisely. Grey and Sadoff (2007) conceptualized water safety in relation to human and ecological health, focusing specifically on protection against connected water hazards. In the last few decades, there have been endeavors to quantify water security by employing specific indicators. The availability of water based on stress due to water and scarcity. While Dragoni et al. (2015) evaluated the worldwide risks to population water availability and riverbank diversity through four distinct categories of stressors: watershed disruption contamination, water supply advancement, and biotic variables. The water security of all Asian countries was assessed by ADB (2016) using five dimensions: residential water safety, socioeconomic water assurances, water supply in cities security, natural water protection, and resilience to water-related disasters.

10.6 The Impact of Changing Climate on Aquatic Systems

10.6.1 Modified Precipitation Patterns

Since 1901, the average annual precipitation in the average rate of growth in the U.S and throughout the world has been 0.03 inches each decade. Nevertheless, certain regions, such as the Southwest, have experienced a more substantial increase in precipitation. Surface water accessibility for drinking, irrigation, and industry, as well as river drowning and the survival of animals and vegetation, can be significantly impacted by precipitation, which can also have a significant effect on humanity and biodiversity. Changes in precipitation can disrupt natural processes, particularly when they occur at a rate that surpasses the capability of species of both plants and animals to adapt. The increase in Earth's surface temperatures results in increased evaporation and transpiration, which in turn adds more moisture to the air, thereby increasing the overall precipitation. It is anticipated that precipitation will rise in numerous regions as a result of a warming climate. Nevertheless, climate change also impacts precipitation by altering wind patterns and ocean currents, which leads to a decrease in precipitation in certain areas. Furthermore, regions that experience an increase in precipitation may not have an abundance of water availability for ecosystems and humans as a result of the increased evaporation (US EPA, 2024b).

10.6.2 Climate Change Accessibility of Fresh Water

The sustainability of development, biodiversity, and access to water and sanitation are all at risk due to the increasing scarcity of water, pollution, and the unpredictable nature of extreme weather events. Infrastructure can be damaged by flooding and rising sea levels, which can contaminate land and water resources. Meltwater is supplying river systems, while glaciers, ice peaks, and snow fields are disappearing. The regulation of freshwater resources in lowland areas is influenced by volatility in the cryosphere. Civil unrest and migration are precipitated by droughts and wildfires, which destabilize communities. Water scarcity and food insecurity are exacerbated by deforestation, which also diminishes groundwater recharge and exacerbates soil erosion. The degradation of water-dependent carbon sinks is exacerbated by the increasing demand for water, which necessitates energy-intensive water extraction, transportation, and treatment. The scarcity of water can be further exacerbated by water-intensive cultivation for the production of biofuels and sustenance. Climate change threatens the accessibility of freshwater, with just 0.5% of Earth's water being useful and available. Over the past two decades, terrestrial water storage has diminished by 1 cm annually, impacting water security. By 2050, the population in danger of flooding will rise from 1.2 billion to 1.6 billion, while 2.7–3.2 billion individuals will reside in water-scarce regions. More than one-fifth of the world's basins have undergone fast expansions in surface water areas, signifying either flooding or desiccation of water bodies. The ambition of new climate change mitigation promises for 2030 must increase fourfold to restrict global warming to 2 °C and sevenfold to constrain it to 1.5 °C. The Arctic sea-ice extent is at its lowest since 1850 and is anticipated to see ice-free conditions at least once before 2050. Sustainable water management will not be realized until 2049, since 40% of nations continue to possess the ability to reconcile conflicting needs and address escalating challenges, including climate change (Water security under climate change, 2021).

10.6.3 Competition in Water Resource Accessibility

Water supplies are critical for biodiversity as well as for social and economic development. Rise in temperature due to environmental change have a considerable impact on water supplies and will have a compounding effect over time. This includes rising water and energy consumption, variation in industrial activity and increased pollution. Multiple studies show that the influence climatic change conjecture on hydrology in many parts of the world (Jose & Dwarakish, 2020), but a few investigations of Asian catchments exist. Many hydrological and ecohydrological models are physical process based (Sahoo et al., 2018). Climate change is anticipated to affect groundwater supplies as a result of variations in precipitation, evapotranspiration, and hydrological balance. Elevated precipitation results in augmented water flow, heightened flood danger, and diminished groundwater drainage. Elevated temperatures lead to increased respiration and evaporation, hence augmenting the need for irrigation water. To address future difficulties, the effects on water levels due to climatic change must be assessed from the regional to the local scale. Several studies have progressively addressed this issue, employing general circulation models to comprehend climate change and forecast future climatic situations. This choice is essential for regional- or local-scale assessments. A dynamic descent utilizing a regional climate model (RCM) generates almost steady climatic change; nonetheless, it remains susceptible to systematic mistakes. In RCM models, bias correction is frequently applied to examine how climate effects hydrology and water resources at the regional and local scales. General circulation models are used in relevant exposure pathways; however, a set of events chosen by climate researchers to estimate likely potential scenarios to evaluate numerous environmental issues (IPCC, 2013).

10.6.4 Floods and Droughts Cycle Disturbance

Intense precipitation shows alarming effect due to global warming since atmospheric water vapor concentrations grow proportionately to temperature. Whereas change in environment dynamics, such as the extension of tropical Hadley circulation, can either reduce or strengthen this influence geographically. Water supply also influences the moisture-temperature connection. The rate at which severe precipitation scales with land surface temperature varies by region, with some areas experiencing negative scaling due to a lack of moisture or dry conditions (Drobinski et al., 2016). In wet places, heightened air moisture convergence can exacerbate the consequences of excessive precipitation, but in dry conditions, precipitation rise may be set-off by evaporation. Recent research indicates that 30-year mean annual precipitation maxima aggregated across arid and wet regions are expected to grow (Donat et al., 2019). Understanding the links between climate change's influence on severe events and water supply is critical for future-proofing global change planning and sustainable socioeconomic development at the regional level. Flood features change based on regional distribution, temporal evolution, precipitation rarity, antecedent soil moisture conditions, snowmelt timing, and snowpack size.

10.6.5 Disbalance of Ecosystem Equilibria

The fast rate of ecosystem degradation due to human and natural influences has resulted in annual expenditures of billions of US dollars for ecosystem restoration (Zhang et al., 2000). Restoration has become an essential instrument for tackling the biodiversity issue and rehabilitating degraded ecosystems. For decades, ecological restoration efforts have employed several methodologies, mostly focused on facilitating the return of essential ecosystem features. Nonetheless, the discipline of ecological restoration is rather nascent and has yet to properly exploit the opportunity to identify overarching similarities across various restoration initiatives to comprehend ecosystem resilience, recovery, and functionality. Research on the impact of restoration initiatives on ecosystem recovery and rates predominantly focuses on research that observes individual sites over brief durations. The absence of data on overarching recovery patterns within ecosystems complicates the evaluation of theoretical frameworks and solutions (Meli et al., 2014).

Climate change increases the likelihood of extreme dry and wet occurrences on land, as well as precipitation variability. Prolonged heating and the accumulation of water in the atmosphere further enhance ecosystem flammability, as well as physiological and hydrological stress. The physiological stress of many creatures and ecosystems in the water is compounded by long-term acidification trends and increased heatwave occurrences. Ecosystems' sensitivity to climate change is worsened by the combination of various environmental stresses such as habitat loss, excessive fishing, exotic plants and animals, disintegration, and direct habitat damage. Predicting variations and possibilities of declining biodiversity is a tough challenge, since it includes both the delicate impacts on certain species in dynamic, multitrophic environments and the additional dramatic consequences of ecosystem degradation. The complexity of ecosystems, as well as a lack of comprehension of the way organisms and relationships among species react to changes in the climate, requires the application of adaptive ecosystem management solutions. As ecosystems experience enormous transitions as a result of worldwide transformation agents, traditional research and modeling frameworks are obsolete owing to developments in the actual world. Innovative frameworks and enhanced tactics are essential for dynamic, complex systems, and adaptive ecosystem management. Prolonged observation is essential for comprehending and alleviating environmental alterations (Brienen et al., 2015).

10.7 Adaptation to Climate Change for the Assurance of Food and Water Security

Climate change adaptation must not be regarded as an isolated endeavor from additional ecological, social, and financial issues that influence the developmental prospects of impoverished populations. In nations where majorly impoverished individuals rely on revenue from agriculture, suggested measures focus on enhancing agricultural output and reducing agriculture's susceptibility to climate-related stressors. Water conservation for agricultural output is a vital element that must adjust to climatic and socioeconomic constraints in the forthcoming years. Alterations in water utilization will be influenced by variations in water supply, demand, urban expansion, industrialization, and water management practices.

In relation to agricultural productivity and water, climate change adaptation may encompass (Bates et al., 2008).

- The implementation of crop varieties and species that demonstrate enhanced resilience to dry weather, shock, and extreme temperature. A public–private collaboration sponsored by the African Agricultural Technology Foundation, known as Water Efficient Maize for Africa (WEMA), aims to cultivate drought-resistant maize varieties in Africa. This initiative faces opposition as it incorporates biotechnology alongside traditional breeding and marker-assisted techniques.
- Modification of irrigating methods involving quantity, scheduling, or technologies (e.g., drip irrigation systems).
- Implementation of water-conserving strategies to collect water, preserve soil hydration (e.g., crop residue retention, zero-tillage), and mitigate sedimentation and salinity incursion.
- Enhanced water regulation to eliminate water retention, weathering, and nutrient leakage.
- Adjustments of harvest schedules, specifically the scheduling or geographic placement of agricultural operations.
- The incorporation of the agricultural, cattle, forestry, and aquaculture industries at both the farm and watershed area levels.
- Execution of periodic weather forecasts.
- Further adaptive approaches may entail land utilization modifications that leverage changed agroclimatic situations.

Water-associated adapting techniques influence the livestock sector, encompassing enhanced pasture rotation, optimized grazing schedules, selection of animal types and species, consolidation of crops and animal husbandry, tailored feed crops, and sufficient water resources. Land users and rural communities are currently adjusting to political, financial, social, ecological, and climatic transformations. Nonetheless, certain aspects of this adaptation may be maladaptive, including deforestation for agricultural purposes, intensifying agriculture on marginal lands, implementing unsustainable methods, importing non-native flora and fauna, or increasing chemical inputs, resulting in pollution. This may exacerbate land destruction and threaten biological diversity, thereby diminishing the capacity to address climate risks in the future.

According to the IPCC, one way to alleviate the impacts of climatic alterations and boost ecologically sound agriculture output is for agricultural production systems to widely use adaptation measures. Nevertheless, because of diverse regions, lengthy decision-making processes, possible obstacles to change, and implementation delays, the efficacy of these tactics is little known. Less than 10% of the promised money from wealthy nations has really arrived in developing nations, and government assistance for impoverished smallholders is severely constrained. Coordination of water usage across sectors and transboundary river basins, as well as arbitration in the growing number of water disputes, will be major policy priorities in the coming years. Other needs, such as industrial production and urban consumption, would have increasing difficulty in meeting the rising demand for agricultural water due to a decline in precipitation. Conflict is more likely to arise as a result of transboundary framework agreements in the forthcoming years resulting from climate change and rising agricultural water demands. Increasing storage capacity, building in irrigation infrastructure, and modifying land tenure are all unilateral actions that might enhance conflict for water resources due to change in climate and its effects on water accessibility. Thus, better management at both local and worldwide scales is necessary to address the unintended consequences of increased upland water consumption, which hinders downstream consumers of the fluid required for survival (Table 10.1).

10.8 Future Directions

The adaptation to climate change is a substantial societal challenge, as it addresses climate justice issues and the uncertainty of future impacts.

10.8.1 Indigenous Knowledge of Climate Change Adaptation

The traditional wisdom and practices of native peoples are recognized as crucial resources for climate change mitigation; however, they have frequently been neglected in policy and investigation. The IPCC (2018) Special Report on Global Warming of 1.5 °C underscores the difficulty of restricting the increase in temperature to less hazardous thresholds. Global temperatures could still increase by more than 3 °C above pre-industrial levels, even if all signatories to the Paris Agreement fulfill their centrally destined contributions. The United Nations Conference of Parties in 2021 will conduct a review of progress made and necessary measures to prevent a climate calamity. Adapting to the effects of changing climates will become increasingly critical, particularly for marginalized and vulnerable populations. Nevertheless, the adaptation gap report by the United Nations Environment Programme (2022)

Table 10.1 The relationship between adaptation strategies, their impact on food and water security, the obstacles they present, and the sources that support these insights

Adaptation strategies	Food security impact	Water security impact	Challenges	References
Practices of sustainable agriculture	Augments resistance to climatic variability and improves agricultural productivity	Efficient irrigation reduces water consumption	Funds are required to support education and technology	FAO (2015)
Diversifying the crops	Decreases dependence on commodities that are susceptible to climate change	Enhances water efficiency and facilitates water retention	Potential for market volatility and untested agricultural outputs	Revi et al, (2014), FAO (2018)
Methods for water conservation	It sustains the long-term agricultural water supply	Enhances groundwater replenishment and minimizes waste	Significantly high implementation costs and the need to alter behavior	World Bank (2020)
Drought-resilient crop varieties	Facilitates food production in dry environments	Reduced dependence on water resources that are excessive	Limited access to seedlings and training is in place	FAO (2015)
Conservation of rainwater	Provides additional watering during dry months	Enhances the availability of water for both agriculture and households	Expenditures on maintenance and infrastructure	IPCC (2022)
Agricultural reforestation	Improves soil fertility and mitigates erosion	Increases water retention and protects water supplies	Land tenure concerns and the necessity of long-term planning are essential	Quandt et al. (2023)
Enhanced water management	Enhances the efficacy of food production	Protects water resources and encourages equal distribution	Necessitates infrastructure and improvements in governance	IPCC (2022)
Technological innovations (e.g., artificial intelligence, Internet of things)	Improves precision agriculture, minimizes waste	Enhances water use through real-time data analysis	Expensive and necessitates technical proficiency	Neufeld (2020), IPCC (2021)

reveals a deteriorating disparity between the preparedness of countries for climate adaptation and the necessary actions to equip people for a period characterized by escalating climate threats (Petzold et al., 2020).

10.8.2 Equitable Resolutions

Climate change is a worldwide issue that impacts every area of existence, with developing countries anticipated to face severe repercussions. The necessity of equitable solutions to mitigate social, economic, and environmental consequences is emphasized by the discrepancy in adaptive capacity among nations. Globally, algorithms and innovations are being researched to improve climate mitigation, while emerging nations prioritize climate justice and vulnerability reduction.

10.8.3 Technical Advancements

The Bonn UN Climate Change Conference emphasized the necessity of a holistic strategy to confront the climate catastrophe by addressing the interconnections among water, energy, and food. Agricultural and food systems account for approximately one-third of world energy usage and two-thirds of global freshwater extraction. Family farmers provide 80% of global food production, underscoring the importance of empowering them to assume leadership in climate solutions. Innovation and technology are essential for resolving this challenge. The FAO Hand-in-Hand Approach and the FAO Technological Rural Communities Initiative demonstrate the influence of technological advances on enhancing productivity, effectiveness, and sustainable resource management. The FAO's Climate Change Policy and Innovative Science and Technological Strategies are essential for enhancing creative solutions in practice. Innovation and technology are crucial instruments for nations to accelerate the execution of their nationally determined contributions (NDCs) and national adaptation plans (NAPs). Coordinated efforts across several institutions and scales are essential for systemic transformation of the water–energy–food nexus. Solar energy is frequently associated with various NDC and NAP initiatives in several nations, facilitating local people to realize cost reductions, diminish CO2 emissions, and enhance agricultural output.

Water-efficient agricultural technologies amalgamate conventional and novel methods, technology, and services pertinent to a particular locale to adjust to changing temperatures and unpredictability. With meticulous design and execution, these technologies possess substantial capacity to alleviate the impacts of climate change on water supplies (CIAT, 2014). A meta-analysis indicated that farm-level modifications might enhance agricultural productivity by approximately 7–15% and conserve water by 25–50% relative to scenarios with no adaptations (Challinor et al., 2014).

Conventional management systems integrated with agroecological strategies (biodiversity, soil maintenance, and water conservation) may represent the only feasible and resilient approach to enhance productiveness, environmental stewardship, and endurance of agricultural output under projected climatic conditions. Effective agricultural managerial procedures and technology, including low-tillage crop cultivation techniques, fertilizer and water distribution optimization, and remnants incorporation, can augment the productivity of crops, enhance water and nutritional efficacy, and reduce greenhouse gas emissions. Water-efficient technologies, including micro-irrigation, rainwater harvesting systems, wastewater reuse, furrow-irrigated raised beds, cover cropping, partial root drying, laser land leveling deficit irrigation, greenhouses, and drainage management, can assist farmers in mitigating the effects of the changing climate and fluctuation. The degree of adoption of smart water technologies is influenced by elements like the socioeconomic characteristics of farmers, the biophysical environment of a specific region, and the features of the new technology. Water shortages may be mitigated with intelligent water management strategies and contemporary irrigation technology, employing localized methods for both supply and demand control (Cutter et al., 2012).

10.8.4 An Emphasis on Social Integration and Gender Equity

There are numerous obstacles to climate change adaptation, such as institutional and governance constraints, information and knowledge deficits, and limited access to finance. The current availability of public sources is insufficient to meet the mitigation financing requirements of emerging nations, which are 10–18 times greater. This is essential for empowering communities on the frontlines of climate change and encouraging investment in adaptation solutions. Localized risk assessments are frequently absent, and accurate climate data is not readily accessible. Furthermore, there is fragmentation in the monitoring, learning, and evaluation of adaptation systems. These issues impede effective planning and decision-making for governmental entities, communities, and the commercial domain. In numerous countries, effective adaptation is also impeded by institutions and administrative limitations, including a deficiency of specialized knowledge and coordination among sectors (Nalau & Verrall, 2021).

In terms of adaptation funding requirements and flows, the AGR 2023 study stresses the absence of social inclusion and gender parity. For promoting social integration and gender equality, hardly 20% of National Adaptation Plans and NDCs allot a particular budget; on average, just 2%. While a quarter of adaptation funding from foreign sources is either gender specific or integrative, hardly two percent is deemed gender responsive (UN EP, 2022). Various forms of inequality, including those based on gender, socioeconomic status, the erasure of cultural and conventional expertise, and the persistence of social stereotypes, are exacerbated by changing climatic conditions. The negative effects of these changes will hit the most vulnerable communities the hardest. To ensure the success of locally led adaptation, it is crucial to comprehend the elements associated with power and justice. Women endure a continual power disparity and face the weight of discrimination and gender inequality, which provide substantial challenges to adaptability. Supporting more fair and effective adaptation requires climate funds that are sensitive to gender, money for adaptation, social inclusion, and better integration of equal rights for women and advancement of women within global climate funds.

10.8.5 Organic Facilities

Nature-based solutions (NBS) denote strategies that utilize ecosystems and natural processes to enhance human welfare and animal populations. By supporting wild species and ecosystems, they provide mitigation and adaptation measures that facilitate the attainment of sustainable development goals. NBS encompasses more than climate change; it often highlights strategies to mitigate its effects. For a plan to be considered viable, it must simultaneously achieve benefits for both humans and the environment (Parmesan, 2023). Ecosystems adjust their reactions to climate change based on their resilience. Human involvement may enhance ecosystems by reinforcing linkages, promoting species mobility, and applying scientific knowledge to preserve coral reefs amid changing climate. Conserving and rehabilitating natural and semi-natural areas enhances the adaptability of ecosystems.

Similarly, several ecosystem-based adaptations and nature-based solutions facilitate ecological resilience, certain human actions that foster adaptation also aid human resilience. To mitigate the risk of catastrophic fires and their associated threats to human safety, reinstating natural fire regimes might expand river systems for enhanced water retention, create more grazing grounds for livestock, and facilitate tree planting. The restoration of ecosystems and agricultural productivity do not consistently align.

10.8.6 Regulations for Monitoring and Evaluation

The past few years have witnessed substantial expenditures in climate change adaptation methods, accompanied by progress in modeling, impact assessment, and comprehension of risks to societal and natural systems (IPCC, 2022). Nonetheless, the surveillance and assessment of these efforts has been insufficient, especially in the Andes area (Llambi and Garces, 2021). The four Andean nations, sharing the same historical context and topographical features, are significantly susceptible to climate change while simultaneously fostering resilience in their economies and ecosystems. Climate change is impacting the availability, accessibility, and utilization of freshwater resources, as increasing water demand and glacial retreat influence regional hydrological cycles. The variability of monitoring and evaluation methodologies in the Andean nations illustrates the range of tactics and local circumstances in which adaptation is executed, even within a singular regional effort. Monitoring systems frequently prioritize biophysical components over sociopolitical and institutional factors, neglecting the intricate nature of monitoring systems in the Andes (Vuille et al., 2017).

10.9 Conclusion

Climate variability causes catastrophic weather events such as flooding, drought, and harsh conditions, posing risks to marine life, vegetation, animals, and people. Agricultural water management strategies include rainfall collection and retention, irrigation, and flood control. Climate change resilience highlights water as a critical component in minimizing harmful weather events. Improving water security is critical for addressing climate-related challenges. Food security is dependent on economic development, global commerce, inventories, assistance programs, and socioeconomic implications on agricultural productivity. To summarize, addressing climate change through effective adaptation is critical and provides an opportunity to build more resilient civilizations capable of thriving in the face of difficulties posed by a changing environment. A mix of diverse techniques, active community engagement, and sustained financing will be critical for achieving successful results in dealing with the consequences of climate change.

References

Ajilogba, C. F., & Walker, S. (2021). Climate change adaptation: Implications for food security and nutrition (pp. 735–754). Springer. https://doi.org/10.1007/978-3-030-45106-6_142.

Asian Development Bank. (2016). Asian water development outlook 2016: Strengthening water security in Asia and the Pacific. https://www.adb.org/sites/default/files/publication/189411/awdo-2016.pdf.

Babel, M. S., Shinde, V. R., Sharma, D., & Dang, N. M. (2020). Measuring water security: A vital step for climate change adaptation. *Environmental Research, 185*, Article 109400. https://doi.org/10.1016/j.envres.2020.109400.

Bates, B. C., Kundzewicz, Z. W., Wu, S., & Palutikof, J. P. (Eds). (2008). Climate change.

Brienen, R. J. W., Phillips, O. L., Feldpausch, T. R., Gloor, E., Baker, T. R., Lloyd, J., Lopez-Gonzalez, G., Monteagudo-Mendoza, A., Malhi, Y., Lewis, S. L., Martinez, R. V., Alexiades, M., Dávila, E. Á., Alvarez-Loayza, P., Andrade, A., Aragão, L. E. O. C., Araujo-Murakami, A., Arets, E. J. M. M., Arroyo, L., & Zagt, R. J. (2015). Long-term decline of the Amazon carbon sink. *Nature, 519*(7543), 344–348. https://doi.org/10.1038/nature14283.

Challinor, A. J., Watson, J., Lobell, D. B., Howden, S. M., Smith, D. R., & Chhetri, N. (2014). A meta-analysis of crop yield under climate change and adaptation. *Nature Climate Change, 4*(4), 287–291. https://doi.org/10.1038/nclimate2153.

Chand, K., Thakur, S., & Kumar, S. (2014). Climate change and food security in India: Contemporary concern and issues. *International Journal of Development Research, V4*(2), 359–365.

CIAT. (2014) *Climate-smart agriculture investment prioritization framework*. International Center for Tropical Agriculture (CIAT), Cali, Colombia.

Climate-Smart Agriculture. (2015). Food and Agriculture Organization of the United Nations. https://www.fao.org/climate-smart-agriculture/en/.

Cutter, S., Osman-Elasha, B., Campbell, J., Cheong, S.-M., McCormick, S., Pulwarty, R., Supratid, S., & Ziervogel, G. (2012). Managing the risks from climate extremes at the local level. In *Managing the risks of extreme events and disasters to advance climate change adaptation* (pp. 291–338). Cambridge University Press. https://www.ipcc.ch/site/assets/uploads/2018/03/SREX-Chap5_FINAL-1.pdf.

Dev, M. S., & Sharma, A. N. (2010). *Food security in India: Performances, challenges and policies*, vol. 7, pp. 1–46. Oxfam India Working Paper.

Donat, M. G., Angélil, O., & Ukkola, A. M. (2019). Intensification of precipitation extremes in the world's humid and water-limited regions. *Environmental Research Letters, 14*(6), Article 065003. https://doi.org/10.1088/1748-9326/ab1c8e.

Dragoni, V. U., Giontella, C., Melillo, M., Cambi, C., Matteo, L. D., & Valigi, D. (2015). Possible response of two water systems in central Italy to climatic changes.

Drobinski, P., Alonzo, B., Bastin, S., Da Silva, N., & Muller, C. (2016). Scaling of precipitation extremes with temperature in the French Mediterranean region: What explains the hook shape? *Journal of Geophysical Research Atmospheres, 121*(7), 3100–3119. https://doi.org/10.1002/2015jd023497.

Grey, D., & Sadoff, C. W. (2007). Sink or Swim? Water security for growth and development. *Water Policy, 9*(6), 545–557.

Hansen, J., et al. (2010). Global surface temperature change. *Reviews of Geophysics, 48.*

IPCC Stocker, T. F., Qin, D., Plattner, G. K., Tignor, M., Allen, S. K., Boschung, J., Nauels, A., Xia, Y., Bex, V., & Midgley, P. M. (2013). Climate Change 2013: The Physical Science Basis. Contribution of Working Group I to the Fifth Assessment Report of the Intergovernmental Panel on Climate Change. Cambridge University Press, Cambridge, 1552 ISBN: 978-1-107-05799-1.

IPCC. (2018). Global warming of 1.5 °C. An IPCC special report on the impacts of global warming of 1.5 °C above pre-industrial levels and related global greenhouse gas emission pathways, in the context of strengthening the global response to the threat of climate change, sustainable development, and efforts to eradicate poverty. In V. Masson-Delmotte et al., (Ed.), *Summary for Policymakers*. World Meteorological Organization.

IPCC. (2022). *Climate Change 2022: Impacts, adaptation, and vulnerability. contribution of working Group II to the sixth assessment report of the intergovernmental panel on climate change*. Cambridge University Press.

Jose, D. M., & Dwarakish, G. S. (2020). Uncertainties in predicting impacts of climate change on hydrology in basin scale: a review. *Arabian Journal of Geosciences, 13*(19). https://doi.org/10.1007/s12517-020-06071-6.

Llambí, L. D., Garcés, A., & Consortium for the sustainable Development of the Andean Ecoregion—Condesan. (2021). Adaptation to climate change in the Andes. CONDESAN. https://weadapt.org/wp-content/uploads/2023/05/adaptation-to-climate-change-in-the-andes.pdf.

Lloyd, S. J., Kovats, R. S., & Chalabi, Z. (2011). Climate change, crop yields, and undernutrition: Development of a model to quantify the impact of climate scenarios on child undernutrition. *Environmental Health Perspectives, 119*(12), 1817–1823. https://doi.org/10.1289/ehp.1003311.

Medek, D. E., Schwartz, J., & Myers, S. S. (2017). Estimated effects of future atmospheric CO2 concentrations on protein intake and the risk of protein deficiency by country and region. *Environmental Health Perspectives, 125*(8). https://doi.org/10.1289/ehp41.

Meli, P., Benayas, J. M. R., Balvanera, P., & Ramos, M. M. (2014). Restoration enhances wetland biodiversity and ecosystem service supply, but results are context-dependent: A meta-analysis. *PLoS ONE, 9*(4), Article e93507. https://doi.org/10.1371/journal.pone.0093507.

Nalau, J., & Verrall, B. (2021). Mapping the evolution and current trends in climate change adaptation science. *Climate Risk Management, 32*, Article 100290. https://doi.org/10.1016/j.crm.2021.100290.

NASA Earth Observatory. (2022). World of change: Global temperatures. https://earthobservatory.nasa.gov/world-of-change/global-temperatures.

Neufeld, J. L. (2020). The state of food and agriculture 2020. FAO. https://doi.org/10.4060/cb1447en.

Opschoor, J. B. (2008). Fighting climate change—human solidarity in a divided world. *Development and Change, 39*(6), 1193–1202. https://doi.org/10.1111/j.1467-7660.2008.00515.x.

Parmesan, C. (2023). Terrestrial and freshwater ecosystems and their services (pp. 197–378). Cambridge University Press. https://doi.org/10.1017/9781009325844.004.

Petzold, J., Andrews, N., Ford, J. D., Hedemann, C., & Postigo, J. C. (2020). Indigenous knowledge on climate change adaptation: A global evidence map of academic literature. *Environmental Research Letters, 15*(11), Article 113007. https://doi.org/10.1088/1748-9326/abb330.

Quandt, A., Neufeldt, H., & Gorman, K. (2023). Climate change adaptation through agroforestry: Opportunities and gaps. *Current Opinion in Environmental Sustainability, 60*, Article 101244. https://doi.org/10.1016/j.cosust.2022.101244.

Ramachandran, N. (2014). Persisting Undernutrition in India. *In Springer eBooks.* https://doi.org/10.1007/978-81-322-1832-6.

Revi, A., Satterthwaite, D., Aragón-Durand, F., Corfee-Morlot, J., Kiunsi, R. B. R., Pelling, M., Roberts, D., Solecki, W., Gajjar, S. P., & Sverdlik, A. (2014). Towards transformative adaptation in cities: The IPCC's fifth assessment. *Environment and Urbanization, 26*(1), 11–28. https://doi.org/10.1177/0956247814523539.

Sage, C. (2014). *Impacts of climate change on food accessibility* (pp. 709–715). Springer. https://doi.org/10.1007/978-94-007-5784-4_120.

Sahoo, S., Dhar, A., Debsarkar, A., & Kar, A. (2018). Impact of water demand on hydrological regime under climate and LULC change scenarios. *Environmental Earth Sciences, 77*(9). https://doi.org/10.1007/s12665-018-7531-2.

The State of Food Security and Nutrition in the World. (2018). *Building climate resilience for food security and nutrition.* Policy Support and Governance. Food and Agriculture Organization of the United Nations. (2018). https://www.fao.org/policy-support/tools-and-publications/resources-details/en/c/1152267/.

Turral, H., Burke, J., & Faures, J. (2011). *Climate change, water and food security*, no 36. Food and Agriculture Organization of the United Nations eBooks. http://agris.fao.org/agris-search/search.do?recordID=XF2006446681.

Ueyama, M., Ichii, K., Kobayashi, H., Kumagai, T., Beringer, J., Merbold, L., Euskirchen, E. S., Hirano, T., Marchesini, L. B., Baldocchi, D., Saitoh, T. M., Mizoguchi, Y., Ono, K., Kim, J., Varlagin, A., Kang, M., Shimizu, T., Kosugi, Y., Bret-Harte, M. S., & Yasuda, Y. (2020). Inferring CO2 fertilization effect based on global monitoring land-atmosphere exchange with a theoretical model. *Environmental Research Letters, 15*(8), Article 084009. https://doi.org/10.1088/1748-9326/ab79e5.

United Nations Environment Programme. Adaptation Gap Report 2022: Too Little, Too Slow—Climate Adaptation Failure Puts World at Risk. Nairobi. 2022. Available online: https://www.unep.org/adaptation-gap-report2022.

US EPA. (2024a). *Climate Adaptation and Source Water Impacts.* US EPA. https://www.epa.gov/arc-x/climate-adaptation-and-source-water-impacts.

US EPA. (2024b). *Climate Change Indicators: U.S. and Global Precipitation.* US EPA. https://www.epa.gov/climate-indicators/climate-change-indicators-us-and-global-precipitation.

Vuille, M., Carey, M., Huggel, C., Buytaert, W., Rabatel, A., Jacobsen, D., Soruco, A., Villacis, M., Yarleque, C., Timm, O. E., Condom, T., Salzmann, N., & Sicart, J. (2017). Rapid decline of snow and ice in the tropical Andes—impacts, uncertainties and challenges ahead. *Earth-Science Reviews, 176*, 195–213. https://doi.org/10.1016/j.earscirev.2017.09.019.

Water Security Under Climate Change. (2021). *Water resources development and management.* https://doi.org/10.1007/978-981-16-5493-0.

World Bank. (2020). Water in Agriculture: Towards Sustainable Agriculture. https://documents.worldbank.org/en/publication/documents-reports/documentdetail/875921614166983369/water-in-agriculture-towards-sustainable-agriculture.

Zhang, P., Shao, G., Zhao, G., Master, D. C. L., Parker, G. R., Dunning, J. B., & Li, Q. (2000). China's forest policy for the 21st century. *Science, 288*(5474), 2135–2136. https://doi.org/10.1126/science.288.5474.2135.

Chapter 11
Outcomes, Trends, and Challenges of Food Systems in a Globalized World

Mukul Machhindra Barwant, Rachna Nagarkar, and Yogesh Ugale

Abstract In the context of a rapidly globalizing world, food systems are undergoing significant transformations that present both opportunities and challenges. This chapter explores the multifaceted outcomes of these changes, emphasizing the critical issues of food security, sustainability, and environmental impact. The global food trade has become a pivotal mechanism for addressing food shortages, particularly in regions facing high food insecurity; however, it is not a panacea for the looming global food crisis exacerbated by population growth and dietary shifts toward more resource-intensive foods. The sustainability of food systems has emerged as a pressing concern, especially in light of climate change and the increasing frequency of natural disasters. The Food and Agriculture Organization (FAO) defines a sustainable food system as one that ensures food security and nutrition for all while safeguarding the economic, social, and environmental foundations necessary for future generations. Yet, the current trajectory of food systems often undermines these principles, leading to adverse outcomes such as biodiversity loss, greenhouse gas emissions, and the exacerbation of malnutrition. For instance, the fragility of the global food system has been highlighted by recent disruptions caused by the COVID-19 pandemic and conflicts such as the Russia–Ukraine War, which have exacerbated food insecurity and malnutrition. Moreover, the complexity of global food systems has intensified, driven by advancements in technology and globalization. While this complexity can enhance resilience by diversifying food sources, it also poses risks of over-complexification, which can lead to environmental degradation and increased vulnerability to shocks.

Keywords Food systems · Food production · Sustainability · Globalization · Nutrition

M. M. Barwant (✉) · Y. Ugale
Department of Botany, Sanjini Rural Education Society, Sanjivani Arts Commerce and Science College, Kopargaon, Maharashtra, India
e-mail: mukulbarwant97@gmail.com

R. Nagarkar
Department of Microbiology, Sanjini Rural Education Society, Sanjivani Arts Commerce and Science College, Kopargaon, Maharashtra, India

J. Parray and A. K. Haghi (eds.), *Food and Water Security*,
https://doi.org/10.1007/978-3-032-04870-7_11

11.1 Introduction

A complex and ever-changing network, the global food system includes the production, distribution, processing, and consumption of food. Globalization has a significant impact on food systems all around the world as it continues to reshape economies and cultures. Policymakers, companies, and consumers must all recognize the potential and challenges presented by this interconnectedness. Traditionally, food systems have been thought of as a collection of processes that start with production and end with consumption. Food security, however, is a complicated problem with a wide range of social, political, environmental, and economic factors. It includes elements related to accessibility, usability, and availability. It is necessary to define a "food system" more broadly in order to conduct a thorough and comprehensive analysis of how the current setup of food production, processing, distribution, and consumption contributes to food security. Food security is also impacted by a wide range of additional economic, social, and environmental factors, and their complicated relationships (Ericksen, 2008).

Food systems have historically changed from small-scale, subsistence-based models to large-scale, internationally linked systems fueled by trade and scientific discoveries. Food production and consumption have changed due to the emergence of multinational agribusinesses, advancements in logistics and transportation, and the expansion of global eating patterns (Béné et al., 2019). Although these modifications have made it easier to obtain a wide variety of foods, they have also added new intricacies and weaknesses to the food chain (FAO, 2020). Food security is when people have physical and financial access to enough safe, nourishing food to meet their dietary needs and preferences for an active and healthy life. If food systems are understood broadly and generically, then the status of food security for any group can be seen as the main result of these systems. Globally, efforts to improve nutrition and lower the incidence of hunger have been successful due to increases in the productivity and efficiency of food systems. These achievements, however, are overshadowed by grave worries about those facets of food systems that jeopardize social, economic, and environmental objectives, hence jeopardizing food security. Additionally, given the social, political, and environmental circumstances in the context of social, political, and economic changes may bring unprecedented stresses to bear on food systems and food security (Ericksen, 2008).

Globalization has its advantages, but the food system still has a lot of problems. These include social challenges like labor exploitation and unequal access to resources, as well as environmental sustainability concerns such as soil degradation and water scarcity (Godfray et al., 2010). Furthermore, the COVID-19 pandemic has brought attention to the vulnerability of international supply networks and the requirement for food systems to be resilient (FAO, 2021). Examining food systems in a globalized world entails assessing the effects of technological advancements on food production and distribution, comprehending the influence of international trade policies, and analyzing the interdependencies between local and global food networks (Hertel, 2018). It also entails investigating the ways in which global trends—like dietary modifications and climate change—affect sustainability and food security. Future food systems must be more resilient and flexible in order to handle sustainability issues and endure shocks to the world. This entails creating cutting-edge food production technology, improving supply chain transparency, and putting laws into place that support fair access to wholesome food (Smith et al., 2019). Food system foresight studies are paying more attention to nutrition and diets because these aspects have significant effects on the environment and human health. The symbiotic relationships among diet, foodborne illnesses, and health mostly impact certain farmers, notably those living in rural areas. There are, in fact, both direct and indirect relationships unique to the agricultural industry that need to be taken into account. The first set of interactions focuses on the effects of nutrition and food consumption on health, while the second set is concerned with the influence of agricultural earnings on the availability of food and medical care. The living conditions of rural impoverished people are made worse by under-nutrition and poor diets, which result in ill health and an incapacity to work (Maggio et al., 2019).

The twenty-first century is essentially marked by social and economic change in both food systems and food security. Examples of these changes include the notable intensification of food production, the phenomenal growth of food product processing and packaging, corporate concentration in retailing and distribution, and the growing influence of large numbers of urban consumers. Creating policies to guarantee food security is a very difficult task that calls for a thorough and well-rounded analytical approach (Ericksen, 2008). Examining food systems in a globalized world entails assessing the effects of technological advancements on food production and distribution, comprehending the influence of international trade policies, and analyzing the interdependencies between local and global food networks (Hertel, 2018). It also entails investigating the ways in which global trends—like dietary modifications and climate change—affect sustainability and food security.

Future food systems must be more resilient and flexible in order to handle sustainability issues and endure shocks to the world. This entails creating cutting-edge food production technology, improving supply chain transparency, and putting laws into place that support fair access to wholesome food (Smith et al., 2019). A wide range of outcomes pertaining to human and environmental health and well-being, as well as opportunities to promote fairness and sustainability along food value chains, can be aided or hindered by actors in modern food systems. Simultaneously, a core group of interconnected challenges related to food systems have been recognized by health professionals, food and agriculture companies, environmental and poverty activists, and policymakers. These challenges include how to sustainably produce enough food, how to help those who grow it make a living, and how to obtain and maintain a healthy diet (Jones et al., 2019).

The most common way to examine food security is to look at why people lack it, or why they are undernourished or hungry. Although there are still many different types of food insecure conditions in society, over the twentieth century, the nature of food insecurity underwent a significant change. Nowadays, social causes are seen as being essentially significant. Changes in the bio-geophysical environment, which can be brought about by human activity or natural processes, are included in the category of global environmental change. These alterations could take on global proportions or perhaps start locally but spread so far that they are considered a world-wide phenomena (Ericksen, 2008). The evolution, management, and performance of food systems are greatly impacted by our world's urbanization. In particular, the food system's operations are becoming recognized as a crucial component of the problems that most governments face, including generating more and better jobs, combating climate change and enhancing resilience, guaranteeing food security, and enhancing nutrition and health. Issues related to the food system have traditionally been addressed at the federal, state, and local levels; municipalities and metropolitan districts have not prioritized them to date. This is evolving. Urban food systems are gaining popularity, and local governments are tackling the real-world food issues that their communities and residents face, such as limited access to wholesome food, food insecurity, the need for jobs in the food system, and the large percentage of food that is stored in solid waste (Tefft & Jonasova, 2020).

Better public health, less of an adverse effect on the environment, and more food security are the results of a properly run global food system. The future of food systems is being shaped by contemporary trends including plant-based diets, digital agriculture, and circular economy methods (Garnett, 2013).

Food systems have made it possible to grow enough food to feed the world's rapidly growing population while also reducing the number of devastating famines that killed hundreds of millions of people. However, this great acceleration has brought with it trade-offs and new challenges, especially in relation to climate change, ecosystem resilience, and growing inequality. These obstacles have hampered efforts to guarantee that everyone is well nourished. As a result, the goals of the food system have changed from being to feed everyone with enough nutrient-rich food to do so in an environmentally sustainable manner while also supporting social justice, fair and equitable livelihoods, and respect for cultural values (Ambikapathi et al., 2022). Improving food security without further jeopardizing social and environmental welfare outcomes is a formidable issue. The character and trajectory of numerous shifts in the food system indicate that addressing this challenge will be difficult. Highly interconnected commodity markets impact global food security and the environment. Thus, resolving issues related to food insecurity and loss of ecosystem services requires an awareness of the intricate relationships between various processes. System-based methods are required to assist in transferring this knowledge (Ericksen et al., 2012).

It is evident that if we want to make fairness, sustainability, and health the top priority in food production and consumption, the global food system must be transformed. While new technology will play a part in some of these transitions, others will call for changes in public policy, adjustments made by the business agro-food sector, initiatives from civil society, and individual behavioral adjustments. An overview of alternative food efforts is provided in this chapter. Consumers, public, are spearheading these initiatives and commercial institutions, and non-profit groups, and they are all actively seeking to create good impacts and improvements along food value chains. These initiatives are changing people's perceptions of food in relation to social justice, environmental degradation, and climate change in addition to what they consume (Jones et al., 2019).

11.2 Outcomes of Globalized Food Systems

The food system has increased a great attention in the past decades. It is being transformed at an unprecedented rate as a result of global economic and social change. Figure 11.1 represents the outcomes of global food system categorized into various sections.

11.2.1 Economic Impact on Food Production and Trade

Food production and trade are fundamental to the global food system, with complex interactions between local production and international trade. Research indicates that in most regions worldwide (102 countries), food demand is primarily met through local food supply, particularly in coastal and inland areas. However, 31% of countries or regions achieve food security primarily through import trade, especially island nations and peninsular countries. The relationship between currencies and trade significantly impacts food production and trade. Exchange rate volatility generally has a negative effect on trade flows, though the magnitude depends on factors like hedging instruments, production structure, and economic integration. Currency misalignments can have short-term effects on trade, influenced by firms' pricing strategies and global production networks. Food aid's impact on food crop production remains controversial. Some studies suggest a disincentive effect, while others find mixed results. In Zimbabwe, food aid shipments significantly affected maize production in the commercial agricultural sector but showed no evidence of impact in the communal sector (Mudzingiri & Chidoko, 2014). Similarly, research on Sub-Saharan African countries indicated that food aid had a significant negative impact on agricultural output across exportables, importables, and non-traded goods sectors. Climate change is expected to significantly affect agricultural production and farm businesses. A study in Western Australia projected that current farming systems may decline in profitability under climate change, potentially becoming financially unviable in the long term. However, adaptation strategies, such as diversified farming systems, investment in technology, and crop area adjustments, could help mitigate these impacts (Abrahams et al., 2012).

Given this, the economic impact on food production and trade is multi-faceted, influenced by factors such as local production capacity, international trade dynamics, currency fluctuations, food aid policies, and climate change. Understanding these complex interactions is crucial for developing effective agricultural policies, managing food insecurity risks, and promoting sustainable food systems in a globalized world.

Fig. 11.1 Outcomes of globalized food systems

11.2.2 Global Trade Flows in Agriculture

Global agricultural trade has become increasingly significant, constituting approximately one-tenth of all international trade flows. The evolution of global agricultural trade patterns, focusing on the integration of individual markets, trade flow directions between geographical regions, and supply concentration of commodity markets (Lipsis et al., 2017). An interesting aspect to consider is the role of preferential trading agreements in shaping the global agro-food trading environment. The role of biofuels in global agricultural trade could also be explored, discussing how national support policies and trade measures have shaped demand and trade routes for these products (Dadakas & Tatsi, 2021).

11.2.3 Economic Inequality Among Producers

Globalized food systems have led to significant transformations in agricultural markets, creating both opportunities and challenges for producers. While some have benefited from increased market access and export opportunities, others have been left behind or negatively impacted. The integration of small-scale producers into global supply chains, particularly supermarket chains, has been a complex process with varying outcomes (Aruoma, 2008). This integration often comes with stricter food safety and quality standards, as well as logistical requirements, which can be challenging for smaller producers to meet. Interestingly, the case study of avocado production in Colombia highlights how global food trends can reshape local agricultural practices and livelihoods. While large investors have profited from the avocado boom, small-scale farmers have been impoverished, demonstrating the uneven distribution of benefits from globalized food systems. This example underscores the need to examine not just linear supply chains but also the broader social and economic context in which agriculture operates. In conclusion, globalized food systems have exacerbated economic inequalities among producers, often favoring larger, more resourced operations over small-scale farmers. To address these disparities, policies and investments in food-deficit low-income countries should focus on diversifying production and markets, enhancing competitiveness for local farmers and SMEs, and incorporating externalities in international trade (Berkum, 2021). Additionally, there is a need for more inclusive, competitive, and nutrition-sensitive agri-food markets that ensure fair trade of commodities and consider the long-term food security implications for all producers.

11.2.4 Social Outcomes

Globalized food systems have led to complex social outcomes, both positive and negative, affecting various aspects of society worldwide. The integration of global markets has provided promise for better food supplies and needs, potentially improving food security in many regions. However, this has also led to a paradoxical situation where food insecurity persists despite increased global food production. The nutrition transition, marked by a shift to Western diets dominated by highly processed, sugar-sweetened, and high-caloric foods, has emerged as a global crisis, contributing to the triple burden of under-nutrition, micronutrient deficiency, and over-nutrition (Golden et al., 2021). Interestingly, the impact of globalization on food systems varies across different regions and socio-economic contexts. In low-income countries, the consequences of global events like the COVID-19 pandemic have been more disruptive to short food supply chains compared to high-income countries. This highlights the uneven distribution of benefits and risks associated with globalized food systems. Additionally, the influence of globalization on local food systems is shaped by factors such as socio-historical trajectories, socio-economic and socio-cultural contexts, and ongoing local environmental issues. So the globalized food systems have improved food availability in many areas, they have also contributed to nutritional transitions and health challenges. The social outcomes of these systems are complex and context dependent, emphasizing the need for integrated approaches involving stakeholders from all policy domains associated with food supply chains to ensure sustainability and equitable distribution of benefits (Bushueva et al., 2024).

11.3 Access to Food and Nutrition Security

Globalized food systems have significantly impacted access to food and nutrition security, with both positive and negative outcomes. The increasing interconnections of global food supply chains have improved food availability in many regions, but it has also created new challenges and vulnerabilities. Access to food and nutrition security in globalized food systems is characterized by complex dynamics. While these systems have enhanced efficiency and increased the variety of food available, they have also led to concerns about the nutritional quality of many food items, particularly those high in calories but low in nutritional value. The globalization of food systems has also resulted in a partial convergence in the supply of animal source foods and sugar across countries, while simultaneously causing a divergence in the availability of vegetables, seafood, and oil crops. Interestingly, the impact of globalized food systems on access to food and nutrition security varies significantly across regions. For instance, many Western countries have experienced a decline in the supply of animal source foods and sugar, while countries like South Korea, China, and Taiwan have seen substantial increases in the availability of these food groups along with vegetables, seafood, and oil crops. In contrast, sub-Saharan African countries have shown remarkably little change in food supply patterns over the past five decades. In conclusion, while globalized food systems have improved overall food availability, they have also created new challenges for ensuring equitable access to nutritious food. The increasing reliance on global supply chains has made food systems more vulnerable to economic shocks, price volatility, and disruptions such as the COVID-19 pandemic. To address these challenges and enhance food and nutrition security, there is a need for accelerated investments in more inclusive, sustainable, and resilient food systems that can better withstand future crises and support global food security.

11.4 Effects on Rural and Urban Populations

Globalized food systems have led to significant changes in food availability, dietary patterns, and nutritional outcomes worldwide. The increasing reliance on global supply chains and large-scale distribution systems has not only improved efficiency but also created new challenges. These systems have contributed to the rising burden of malnutrition in all its forms, driven by a global transition toward unhealthy ultra-processed food consumption patterns. Interestingly, while globalized food systems have improved food availability and nutritional diversity, they have also increased the vulnerability of some countries' food systems to global natural disasters and political instability (Wang & Dai, 2021). The intensification of international food trade and globalization processes has implications for population nutritional status worldwide, with potential effects on the prevalence of overweight and obesity (Silvestrini et al., 2023). The outcomes of globalized food systems are complex and multifaceted and have improved efficiency and access to diverse foods, they have also contributed to the nutrition transition and associated health problems. The challenge lies in developing sustainable food systems that provide universal access to healthy food choices for a growing world population while reducing the environmental footprint of the global food system (Lindgren et al., 2018). This requires addressing the interconnected dynamics of socio-economic and natural systems within food systems to avoid social-ecological traps and accelerated nutrition transitions (Golden et al., 2021).

11.5 Environmental Outcomes

Globalized food systems have significant environmental impacts across various stages of the supply chain. The modern food production system is highly resource-intensive, contributing to unsustainable practices that degrade land and water resources, ecosystem services, and biodiversity (Jeswani & Azapagic, 2022). These systems also lead to increased greenhouse gas emissions and deforestation. The transport components of food supply chains, often overlooked, have substantial environmental consequences. The global sourcing of food produce, centralized distribution systems, and car-based shopping have increased the distance between producer and consumer, known as food miles (Jones, 2002). This increase in transportation is responsible for a considerable fraction of the total energy consumption in the life cycle of fresh produce, often exceeding the energy consumed in cultivation. Interestingly, contrary to common belief, long food supply chains may generate less negative environmental impacts than short chains in terms of fossil fuel energy consumption, pollution, and GHG emissions per kg of a given product. This suggests that the environmental impacts of food distribution are not solely determined by geographical distance but depend on numerous factors, including supply chain infrastructure. In conclusion, while globalized food systems offer convenience and year-round availability of fresh produce, they come with significant environmental costs. These impacts occur through land use change, agricultural production, and transportation. To address these issues, there is a growing need for measurement and management of environmental impacts in global food supply chains (Accorsi et al., 2015), and the development of more sustainable alternatives such as optimized food supply networks and local production systems (Jones, 2002).

11.5.1 Land Use Changes and Deforestation

Land use changes and deforestation have significant impacts on globalized food systems, affecting both food production and environmental sustainability. Deforestation, primarily driven by agricultural expansion for food production, is a major contributor to global forest cover changes. Approximately 90% of forest cover changes between 2000 and 2018 were attributable to agricultural expansion (Sylvester et al., 2024). This trend has profound effects on the hydrological cycle, with deforestation decreasing global vapor flows from land by 4% (3000 km^3/yr), comparable to the increased vapor flow caused by irrigation (2600 km^3/yr). Interestingly, while deforestation aims to increase food production, it can have counterproductive effects. The modification of vapor flows due to deforestation, particularly in regions like sub-Saharan Africa and the Asian monsoon region, may affect the capability to increase food production in other areas (Gordon et al., 2005). Moreover, deforestation enriches antibiotic resistance genes (ARGs) in soil, potentially compromising public health through the food chain (Lemos et al., 2020). In conclusion, the relationship between land use changes, deforestation, and globalized food systems is complex and interconnected. While deforestation is often driven by the need to increase food production, it can lead to unintended consequences that threaten long-term food security and environmental sustainability. Addressing these challenges requires a holistic approach that considers the entire food system, including consumption patterns, production methods, and trade dynamics. Sustainable solutions may involve agricultural intensification, land use zoning, forest protection, and harnessing globalization to increase land use efficiency rather than uncontrolled expansion (Lambin & Meyfroidt, 2011).

11.5.2 Water Usage and Pollution

Globalized food systems are significantly affected by water usage and pollution, with far-reaching consequences for food security, environmental sustainability, and human health. Water pollution, particularly from agricultural runoff containing nitrates and heavy metals, poses a serious threat to food systems. Nitrate leaching is a primary source of groundwater contamination and freshwater ecosystem degradation. This pollution can impact crop yields, soil quality, and ultimately food safety. Heavy metal concentrations in drinking water often exceed international guideline values, leading to chronic poisoning and adverse health effects through food chain contamination. Climate change exacerbates these issues by altering precipitation patterns and increasing the frequency of extreme weather events like droughts and floods. This affects crop yields, nutritional quality, and overall food production (Phu, 2021). The global food supply chain is vulnerable to these climatic disruptions, potentially leading to food insecurity and economic instability. In conclusion, addressing water usage and pollution in globalized food systems requires a multifaceted approach. This includes improving agricultural practices to reduce nitrate leaching, implementing sustainable water management strategies, and developing robust technologies to enhance drinking water quality. Additionally, integrating blockchain technology into food supply chains could improve traceability and food safety, while adopting sustainable agriculture practices can help mitigate climate-related risks in supply chains.

11.5.3 Carbon Footprint and Climate Change

Globalized food systems are significantly impacted by carbon footprint and climate change, with far-reaching consequences for food security, production, and distribution worldwide. Climate change, driven by greenhouse gas emissions, is altering the physical biosphere and undermining the resilience of natural systems (White & Yeates, 2018). This has led to increased frequency of extreme weather events such as droughts, floods, and cyclones, which threaten food security and clean water availability (A & Parthiban, 2022). Interestingly, while food systems have become increasingly globalized in recent decades, there is a renewed interest in local production and consumption as a means to contribute to sustainable and secure food systems in the face of climate change and urbanization. However, the scale of production can significantly impact environmental health outcomes, with local industrial production sometimes having a lower carbon footprint than small-scale civic production, as demonstrated in a study of chicken meat production in Sydney, Australia. In conclusion, addressing the impact of climate change on food systems requires a multifaceted approach. This includes developing climate-resistant crops, improving water management, and enhancing the resilience of food systems to extreme weather events. Additionally, sustainable agriculture practices can help farmers adapt, maintain, and improve productivity without applying harmful techniques, thus managing and mitigating climate-related risks in their supply chains (Phu, 2021).

11.6 Key Trends in Global Food Systems

11.6.1 Industrialization of Agriculture

The industrialization of agriculture has emerged as a key trend in global food systems, characterized by increased mechanization, use of chemical inputs, and large-scale production methods. This shift has significantly impacted agricultural practices and food production worldwide. Industrialized agriculture has led to greater complexity in food systems since the mid-twentieth century, with innovations such as mechanization, irrigation, genetic modification, and globalized supply chains. While these developments have increased productivity, they have also resulted in environmental degradation and made food systems more dependent on external inputs and vulnerable to disruptions (Loring & Sanyal, 2021). The reliance on chemical fertilizers and pesticides for productivity and pest management has become a major constraint affecting global food production (Mondal, 2022). This dependence poses environmental threats and highlights the need for new innovations in agriculture. Additionally, the industrialization of agriculture has contributed to the loss of fertile land due to insufficient soil management, urbanization, and industrialization, compromising future global food and fiber production (Blum, 2013). The industrialization of agriculture has increased food production to meet growing demands, and it has also created challenges for sustainability and environmental health. To address these issues, there is a growing focus on alternative approaches such as vertical farming, organic farming, and the integration of digital technologies in agriculture (Aliev et al., 2023. These innovations aim to balance productivity with environmental sustainability and food security in the face of global challenges like climate change and population growth.

11.6.2 Growth of Agribusiness and Consolidation

The growth of agribusiness and consolidation has emerged as key trends shaping global food systems in recent years. Developing countries have witnessed significant transformation in their agricultural sectors, with governments and stakeholders strengthening agri-food systems and agribusiness linkages to integrate agriculture into the market economy (Akuriba et al., 2021). This has led to accelerated productivity growth, improved food security, and increased employment and incomes.

Global agribusiness concentration has become a crucial issue for developing countries' food systems. Corporate investments have targeted different parts of the food system at various times, subjugating agriculture and domestic food markets to the needs of global agribusiness. However, in some larger developing countries, national capitalists are the primary force behind emerging urban food systems, with the state playing a key role in their consolidation (Wilkinson, 2009). The agri-food trade system is undergoing rapid changes, with important implications for economic development. These changes include rapid growth and structural shifts in agri-food trade, increased consolidation in food supply chains, proliferation of public and private food standards, high and volatile food prices, and increased vertical coordination in the chains (Maertens & Swinnen, 2015). These trends have significant impacts on developing countries' participation in international agricultural trade and their potential for economic development, income mobility, and poverty reduction in rural areas. In conclusion, the growth of agribusiness and consolidation in global food systems has led to both opportunities and challenges for developing countries. While these trends have contributed to increased productivity and market integration, they have also raised concerns about the concentration of power in the hands of large corporations and the potential impacts on small-scale farmers and local food systems. As the agri-food industry continues to evolve, it will be crucial to balance the benefits of consolidation with the need for inclusive and sustainable development in the agricultural sector.

11.6.3 Technological Advancements

Technological advancements have emerged as a key trend in global food systems, revolutionizing various aspects of the food supply chain from production to consumption. These innovations are addressing critical challenges such as food security, sustainability, and efficiency in the face of population growth, climate change, and evolving consumer preferences. In the agricultural sector, the adoption of Industry 4.0 technologies, including Internet of Things (IoT), Blockchain, Big Data, Information and Communication Technology (ICT), Cloud Computing, and Cyber-Physical Systems, is transforming the agriculture food supply chain (AFSC). These technologies are enhancing traceability, food safety, information system management, and decision-making processes. For instance, blockchain technology is emerging as a game-changer in ensuring food safety and security by improving transparency and secure transactions (Oriekhoe et al., 2023). In food processing, advanced technologies such as high-pressure processing, pulsed electric field processing, and hurdle technology are enhancing microbial safety while preserving nutritional and sensory attributes. Non-thermal technologies like ultraviolet light, cold plasma, and ultrasound are extending shelf life without compromising food quality. These advancements are not only improving food safety and quality but also contributing to sustainability by reducing energy consumption and food waste. The integration of artificial intelligence (AI) and data analytics in the food supply chain is optimizing operations, improving demand forecasting, and enhancing overall efficiency. Additionally, innovative packaging systems are contributing to extended shelf life and reduced food waste, aligning with sustainability goals (Pokharel, 2023).

11.6.3.1 Precision Agriculture

Precision agriculture is emerging as a key trend in global food systems, leveraging advanced technologies to optimize farming practices and address the growing challenges of food security and sustainability. This approach integrates robotics, data science, and artificial intelligence to enhance crop yields, reduce resource consumption, and minimize environmental impact (Veluru, 2024). The core principle of precision agriculture is to apply inputs such as fertilizers and pesticides in the right quantity, at the right time, and in the right place, a practice known as "Site-Specific Management" (Hakkim et al., 2016). This approach aims to maximize productivity and profitability while preserving natural resources like water, air, and soil quality (Maurya et al., 2024). Precision farming employs various technologies, including sensors, transmitters, data systems, and unmanned aerial vehicles (UAVs), to make data-driven decisions and integrate them with improved machinery (Joshi et al., 2023). One of the most significant applications of precision agriculture is in addressing water scarcity. With global warming threatening water sustainability, precision agriculture techniques can significantly reduce water usage in irrigation systems. For instance, a study using deep reinforcement learning in a robotic crop farm prototype demonstrated a 41% reduction in water utilization, along with a 26% increase in crop yield and a 33% decrease in chemical use (Addas et al., 2023). The integration of Internet-of-Things (IoT) technologies in precision agriculture is particularly noteworthy. IoT devices are enhancing farmers' control over crop growth and livestock rearing processes, potentially offering life-saving solutions in the face of looming food crises (Trivedi & Sainkhediya, 2024). Additionally, computer vision and AI technologies are driving a revolution in agriculture, promising increased productivity, economic growth, and solutions to global issues such as food security.

11.6.3.2 Biotechnology and GMOs

Biotechnology and genetically modified organisms (GMOs) have emerged as key trends in global food systems, offering potential solutions to address food security challenges and enhance agricultural productivity. Genetic engineering techniques allow for the alteration of organisms' genetic makeup to produce quicker and more beneficial results, potentially increasing crop yields, improving nutrition, and enhancing drought tolerance (Reddy, 2015). The development and commercialization of GM crops represent the fastest technological revolution in agriculture, with global GM planting areas growing at around 10% annually in the last decade (Falkner, 2008). Interestingly, while GMOs offer numerous potential benefits, there are also concerns about possible adverse health effects, environmental damage, and

Table 11.1 Benefits of the GMO crop or food product (Kramkowska et al., 2013)

Crop/Foods	Benefits	References
Rice	Higher content of B-carotene Higher iron bioavailability	Ostaszewska and Sablik, (2008)
Tomato	Higher content of dry matter Delayed ripening process Aroma intensification Virus resistance	Kumar et al., (2008)
Potato	Higher amylopectin content Cyclodextrin production Resistance to viruses and potato beetle Lower alkaloids content	Sakakibara and Saito, (2006)
Milk (cow, goat, sheep)	Increased tolerance for high temperature Modified casein content Lower lactose content	Kondratowicz et al., (2009)
Transgenic fishes (carp, salmon, trout)	Faster growth rate	Ochocki, (2005)

gene pollution. The debate surrounding GMOs is highly polarized, with advocates urging governments to promote their development and critics demanding precautionary regulation. This controversy highlights the need for balanced regulation that considers both the benefits and risks of each application (Zilberman et al., 2018). Overall, biotechnology and GMOs have the potential to significantly impact global food systems by addressing food security challenges, increasing agricultural productivity, and promoting sustainability. However, their adoption and regulation require careful consideration of potential risks and benefits. As the technology continues to evolve, it is crucial to maintain a dynamic and integrated approach to ensure that the benefits of GMOs are realized while minimizing potential negative impacts on human health and the environment (Anderson et al., 2016). Table 11.1 represents the benefits of the GMO crops and their products.

11.6.3.3 Changing Consumer Preferences

Changing consumer preferences are indeed a key trend shaping global food systems. Based on the provided context, here are several points to consider for inclusion in the book chapter: consumer preferences are evolving rapidly, influenced by factors such as health consciousness, environmental concerns, and technological advancements. These changes have both temporal and spatial dimensions, necessitating an agile agriculture supply chain management (AASCM) to ensure uninterrupted supply from farm to fork (Pandey, 2022). The COVID-19 pandemic has significantly impacted consumer views on food safety and purchasing patterns, heightening concerns about the entire food chain (Krishnamoorthy et al., 2021). This has led to increased interest in food storability and alternative-sourcing options, as evidenced by changes in Google search trends during the pandemic. There's a growing demand for sustainable and ethically produced food, driving the need for transparency and meaningful disclosure in food supply chains (Manning, 2018). This trend is pushing businesses to adopt circular supply chain practices, which can enhance resource efficiency and create new revenue streams (Mittal et al., 2024). The shift toward personalized nutrition and alternative protein sources is gaining momentum, reflecting consumers' increasing focus on health and environmental sustainability. Additionally, the adoption of digital technologies in the food supply chain is enabling better traceability and meeting consumer demands for information about their food's origin and production methods. In conclusion, addressing these changing consumer preferences requires a holistic approach involving innovations in farming practices, food processing technologies, and supply chain management. It also calls for increased collaboration between stakeholders and supportive policy frameworks to facilitate the transition toward more sustainable and consumer-centric food systems (Mittal et al., 2024).

11.6.3.4 Organic and Sustainable Foods

The growing importance of organic and sustainable foods as key trends in global food systems can be addressed through several key points: organic and sustainable foods are increasingly recognized as crucial components in addressing global food security challenges and promoting environmental sustainability (Laureati & Proserpio, 2024). These trends are driven by consumer awareness of health and environmental impacts, as well as the need to mitigate climate change and resource depletion (Alae-Carew et al., 2021). Interestingly, the shift toward sustainable diets involves not only increased consumption of plant-based foods but also the development of alternative protein sources and novel food technologies. This includes the rise of plant-based meat and dairy alternatives, which are gaining popularity, especially among younger generations and higher income groups (Alae-Carew et al., 2021). In conclusion, a comprehensive book chapter on this topic should emphasize the multifaceted nature of sustainable food systems, incorporating aspects of consumer behavior, technological innovations, and policy implications. It should also highlight the need for a balanced approach that considers both plant- and animal-sourced foods in creating nutritionally adequate and environmentally sustainable diets (Smith et al., 2021).

11.6.3.5 Rise of Plant-Based and Alternative Proteins

Plant-based meat alternatives, insects, cultured meat, and other novel protein sources are attracting significant investment and research interest as potential solutions to meet the growing global demand for proteins in a sustainable, ethical, and healthy manner. These alternatives are being promoted for their reported health benefits, lower environmental impact, and improved animal welfare compared to conventional animal-based meat. The global population is projected to reach 9.7 billion in the next 30 years, with an expected 80% increase in meat consumption, making sustainable protein sources crucial for food security (Nadathur et al., 2016). However, it's important to note that the evidence supporting claims of superior nutritional, environmental, and ethical credentials for alternative proteins remains unclear in some cases. Significant research gaps exist, particularly regarding longitudinal evidence on the sustained effects of replacing conventional animal-based proteins with alternative sources. Additionally, challenges persist in achieving desirable sensory properties akin to animal-based meat to promote consumer acceptance (Baune et al., 2022). In conclusion, while alternative proteins offer promising solutions to address global food system challenges, their development and adoption require careful consideration of nutritional adequacy, environmental impact, and consumer acceptance.

11.6.3.6 Global Food Supply Chains and Trade Networks

The global food supply chains and trade networks are increasingly characterized by complexity and interdependence, driven by various socio-economic and environmental factors. As globalization continues to reshape food systems, the reliance on international trade has intensified, with approximately a quarter of all food calories being traded across borders (Davis et al., 2020). This trend is particularly pronounced in low- and middle-income countries (LMICs), where globalization of food trade, including foreign direct investment, has led to shifts in the availability and types of food outlets and consumer choices. The ongoing geopolitical tensions, such as the Russia–Ukraine conflict, have further underscored the vulnerabilities inherent in these global supply chains, as disruptions in major agricultural producers like Ukraine and Russia can have cascading effects on food security worldwide (Jagtap et al., 2022). Moreover, the historical trajectory of food systems reveals a shift from self-sufficiency to trade dependency, highlighting the need for a more equitable distribution of food resources to enhance global food security (Porkka et al., 2013). The complexity of modern food systems, exacerbated by technological advancements and globalization, has led to a phenomenon known as over-complexification, which can render these systems more susceptible to shocks and crises (Loring & Sanyal, 2021). This complexity is not merely a challenge but also a potential avenue for innovation, as stakeholders seek to develop sustainable practices that can mitigate risks associated with food supply disruptions. The implications of these trends extend beyond mere economic considerations; they encompass environmental sustainability and social equity. The food supply chain's efficiency often comes at the cost of environmental degradation, with significant contributions to greenhouse gas emissions and resource depletion (Laso et al., 2018). Therefore, a holistic approach that integrates sustainability into food systems is essential for addressing the dual challenges of food security and environmental stewardship (Bilali et al., 2018). As we move forward, it is imperative to foster resilience within food supply chains through innovative practices, policy reforms, and collaborative frameworks that prioritize sustainability and equity (Blay-Palmer et al., 2019). In conclusion, the evolving landscape of global food supply chains and trade networks necessitates a comprehensive understanding of their complexities and interdependencies. By addressing the vulnerabilities exposed by recent global events and embracing sustainable practices, stakeholders can work toward a more resilient and equitable food system that meets the needs of a growing global population.

11.7 Challenges Facing Food Systems in a Globalized World

11.7.1 Climate Change and Its Impact on Agriculture

The challenges facing food systems in a globalized world are multifaceted, with climate change emerging as a critical factor impacting agricultural productivity and food security. As global temperatures rise and weather patterns become increasingly erratic, agricultural systems are forced to adapt to new realities that threaten crop yields and food supply chains. The necessity to double global agricultural production by 2050 to meet the demands of a growing population, projected to reach 9 billion, further complicates this scenario. Current yield improvements for major crops such as maize, rice, wheat, and soybean are insufficient, averaging only 0.9% to 1.6% per year, which is far below the required growth rate of approximately 2.4% per year to achieve this goal (Xu & Li, 2022). Climate change exacerbates existing agricultural challenges, including water scarcity and soil degradation. Liu et al. highlight that the assessment of agricultural water scarcity must consider both blue and green water availability under future climate scenarios, emphasizing the need for sustainable practices that balance crop production with water conservation (Liu et al., 2022). Furthermore, extreme weather events, such as droughts and floods, can significantly disrupt agricultural productivity, leading to increased food insecurity. The interplay between climate change and agricultural practices necessitates innovative solutions, such as the use of biodegradable radiative cooling mulch to alleviate heat stress and water scarcity (Lin et al., 2023). The complexity of these challenges is compounded by socio-economic factors, including population pressures and changing dietary patterns. The FAO has noted that the demand for food is expected to rise due to population growth and increased consumption, which will intensify competition for land, water, and other resources (Godfray et al., 2010). Additionally, the need for sustainable agricultural practices is underscored by the environmental impacts of conventional farming, which contributes significantly to greenhouse gas emissions and biodiversity loss (Sanders et al., 2021). As such, there is a pressing need for interdisciplinary approaches that integrate agronomic, environmental, and social dimensions to develop resilient agricultural systems capable of withstanding the impacts of climate change (Liu et al., 2023). In response to these challenges, advancements in technology, particularly artificial intelligence (AI), are being explored as potential solutions to enhance agricultural sustainability. AI can facilitate precision farming, optimize resource use, and improve crop management practices, thereby increasing resilience to climate-induced stresses (Sakapaji, 2023). However, the implementation of such technologies must be approached with caution, considering the socio-economic implications and ensuring equitable access for all stakeholders involved in the agricultural sector (Ayed & Hanana, 2021). In conclusion, addressing the challenges facing food systems in a globalized world requires a comprehensive understanding of the interconnections between climate change, agricultural practices, and socio-economic factors. Sustainable intensification of agriculture, supported by innovative technologies and interdisciplinary collaboration, is essential to ensure food security and environmental sustainability in the face of a changing climate.

11.7.2 Food Security and Global Inequality

Food security and global inequality present significant challenges to food systems in a globalized world. The paradoxical nature of contemporary food insecurity persists despite the growing integration of global markets, which promised better food supplies. While technological and institutional innovations have allowed food production to keep pace with population growth, access to food remains a critical issue, particularly for vulnerable populations in developing countries. The current food system, characterized by long food supply chains (LFSC) and globalization, has demonstrated its inability to feed the global population adequately while generating negative ecological, environmental, logistical, and nutritional pressures (Soria-Lopez et al., 2022). This dysfunction is evident in the fact that food insecurity and malnutrition still affect half of the global population, highlighting the urgent need for a transition toward sustainability. To address these challenges, a multifaceted approach is necessary. This includes fostering integration of agro-food sustainability transitions and food security in research, policy, and practice. Strategies should focus on supporting increased food availability and access for the urban poor, promoting healthy diets, facilitating physical activity, and supporting urban agriculture (Ruel et al., 2017).

11.7.3 Loss of Biodiversity and Depletion of Natural Resources

The loss of biodiversity and depletion of natural resources pose significant challenges to global food systems, threatening food security and environmental sustainability. This interconnected issue affects various aspects of food production, distribution, and consumption. Biodiversity loss and natural resource depletion exacerbate global food system challenges, including population growth, changing dietary habits, and climate change vulnerability (Vittuari et al., 2019). These factors put immense pressure on natural capital, leading to resource consumption, environmental pollution, and ecosystem degradation (Sousa et al., 2024). The depletion of natural resources, such as minerals, further compounds the problem by contributing to biodiversity loss and environmental degradation. Interestingly, while biodiversity loss is often viewed as a purely environmental challenge, it is crucial for safeguarding food security. Biodiversity supports agricultural production, mitigates pollution effects, provides livelihood outcomes, and ensures diverse diets to combat hidden hunger. Moreover, biodiversity is essential for resilience against future food security risks and is a precondition for genetic modifications in crop development. To address these challenges, sustainable agricultural practices such as precision agriculture, nanotechnology, biofertilizers, and improved land management systems offer opportunities to enhance food security and environmental sustainability.

11.7.4 Political and Economic Instability

The complex interplay between food security and sociopolitical stability has become increasingly evident in recent years, with global food price spikes in 2008 and 2011 triggering political unrest and toppling governments (Barrett, 2013). This interconnection is further complicated by the globalization of goods and people, the emergence of new economic powers, and sustained urban population growth (Lopez-Gunn et al., 2012). Political instability poses a significant challenge to food systems in a globalized world. The Arab Spring highlighted how socio-economic factors such as inequality, poverty, youth unemployment, and a growing youth population can contribute to political turmoil (Kaphahn & Brennan, 2016). Additionally, uncertainty shocks to global food commodity prices can elicit civil conflict in developing countries, as demonstrated by econometric evidence from 1966 to 1990 (O'Trakoun, 2015). The 2008–09 Global Financial Crisis further emphasized the need to understand the political underpinnings of financial instability and its impact on food systems. In conclusion, addressing political and economic instability in food systems requires a multifaceted approach. This includes implementing policies to reduce export price uncertainty for food commodity exporters (O'Trakoun, 2015); reforming global trade in light of ethical considerations (Lopez-Gunn et al., 2012); and developing evidence-based strategies that are efficient, effective, and sustainable (Bilali & Hassen, 2024). Furthermore, international cooperation is crucial in addressing transnational issues such as foodborne disease surveillance and control. As the world becomes more interconnected, addressing these challenges will be essential for ensuring global food security and stability.

11.7.5 Public Health Challenges

The globalization of food supply chains has led to increased interconnectedness, bringing both opportunities and risks. Food security, safety, and sustainability have emerged as critical issues affecting global public health (Khan, 2017). One of the major challenges is ensuring food security and addressing malnutrition in a world with growing population demands. Climate change, poverty, and conflicts further exacerbate food insecurity and malnutrition issues (Lucero-Prisno et al., 2023). The globalized food system has also introduced new food safety risks, as contaminated food can now spread across wider geographic areas more rapidly (Khan, 2017). Additionally, the rise of antimicrobial resistance (AMR) in the food animal industry, particularly in the poultry sector, poses a significant threat to public health globally. Interestingly, while globalization has increased access to diverse food sources, it has also contributed to the rise of unhealthy diets and obesity in many parts of the world. The convenience food culture and the "wellness" phenomenon have implications for global agri-food systems. Furthermore, the current public health workforce appears unprepared to confront the challenges posed by globalization in the food sector, highlighting the need for improved training and education (Lee et al., 2011). In conclusion, addressing public health challenges in food systems requires a multidisciplinary approach. The One Health concept, which recognizes the interconnectedness of human, animal, and environmental health, is crucial for mitigating risks such as AMR. Moving toward sustainable food value chains is one of the greatest challenges of this century, requiring innovations in food distribution, waste reduction, and consumption patterns (Kusch-Brandt, 2020). To effectively tackle these global issues, there is a need for stronger international cooperation, sustainable funding, and a shift in public health thinking from a singular focus on international health to a more nuanced analysis of global health.

11.7.6 Malnutrition and Obesity

The global food system faces complex challenges, with malnutrition and obesity emerging as significant concerns in both developed and developing countries. The coexistence of under-nutrition, micronutrient deficiencies, and overweight/obesity, known as the "double burden of malnutrition," is particularly prevalent in low- and middle-income countries (Kiosia et al., 2024). This phenomenon is driven by rapid urbanization, population growth, globalization, climate change, and income growth, which are transforming food systems and dietary patterns. Interestingly, the nutrition transition occurring worldwide due to globalized agri-food systems is contributing to both under-nutrition and over-nutrition simultaneously (Lowitt et al., 2018). This paradox is especially evident in urban areas, where changing lifestyles, work patterns, and income factors affect food choices and supply. The COVID-19 pandemic has further highlighted the vulnerabilities in global food systems and their impact on individual agency in food choices (Wells et al., 2021). To address these challenges, a holistic food systems approach is necessary, involving multiple sectors such as agriculture, rural development, and public health. Strategies should focus on improving food utilization, dietary diversity, and quality while also addressing the deeper structural drivers of malnutrition embedded in society. Additionally, global governance systems need to be redesigned to facilitate actions for accelerated reduction of under-nutrition and malnutrition in all its forms. Investing in the elimination of hunger and malnutrition not only fulfills a moral obligation but also makes economic sense, with substantial lifetime economic benefits from reducing child under-nutrition (Fan, 2014).

11.7.7 Food Safety and Contamination Risks

Foodborne diseases pose a significant global public health challenge, with enormous human and economic costs. The globalization of food trade has magnified these challenges by introducing contamination from one country to another, multiplying opportunities for contamination during longer transportation periods, and increasing the impact of contamination as products reach more people over wider areas (Glavin, 2003). The expansion of the food industry has further increased the demand for diverse food products, posing new challenges to food safety due to the production, transportation, and handling of food from various sources and countries. Interestingly, while traditional food safety control systems are increasingly challenged to identify intricate hazards, digital food safety control systems are emerging as a revolutionary shift in food safety methodology. These systems employ modern solutions to monitor food quality in real time, detect a wide range of contaminants, and provide highly predictive assessments of food safety risks. Additionally, blockchain technology offers a promising solution by providing a decentralized, immutable, and transparent ledger system that enables rapid identification of contaminated products and facilitates targeted recalls. In conclusion, addressing food safety challenges in a globalized world requires a multifaceted approach. This includes implementing global standards focused on public health outcomes (Glavin, 2003), leveraging digital technologies for enhanced traceability and risk assessment, and integrating blockchain technology into food supply chains. Furthermore, harmonizing food safety and sustainability standards, fostering international cooperation, and investing in education and training across the food supply chain are essential for building a resilient and sustainable food system capable of supporting global food security while protecting public health and the environment (Chatterjee, 2024).

11.8 Conclusion

When one looks at the outcomes, trends, and challenges in food systems in a globalized society, one finds both significant advancements and significant road blocks. This chapter has investigated the impact of globalization on food systems, showing how food production, distribution, and consumption have reached unprecedented levels of connectivity and efficiency. Advances in agriculture technology and practices have been made possible by this connectedness, which has also made it easier to get a variety of food products. But these benefits do come with a lot of challenging challenges. Environmental sustainability remains a key concern as the demand on natural resources rises and vulnerabilities in the food system worsen due to climate change. Societal problems like unequal access to food are making the global food crisis worse.

References

Abrahams, M., Peek, C., Reynolds, C., Falconer, K. -L., Foster, I., Gool, D., & Gardiner, D. (2012). Developing potential adaptations to climate change for farming systems in Western Australia's Northern Agricultural Region using the economic analysis tool STEP. *Australian Farm Business Management Journal, 9*(1), 9–28. https://doi.org/10.22004/ag.econ.131749

Accorsi, R., Manzini, R., & Versari, L. (2015). Glass versus plastic: Life cycle assessment of extra-virgin olive oil bottles across global supply chains. *Sustainability, 7*(3), 2818–2840. https://doi.org/10.3390/su7032818

Addas, A., Tahir, M., & Ismat, N. (2023). Enhancing precision of crop farming towards smart cities: an application of artificial intelligence. *Sustainability, 16*(1), 355. https://doi.org/10.3390/su16010355

Alae-Carew, C., Green, R., Stewart, C., Cook, B., Dangour, A. D., & Scheelbeek, P. F. D. (2021). The role of plant-based alternative foods in sustainable and healthy food systems: consumption trends in the UK. *Science of the Total Environment, 807*(11), Article 151041. https://doi.org/10.1016/j.scitotenv.2021.151041

Aliev, R., Samoylova, A., & Kurbanova, M. (2023). Transformative potential of digital agriculture for enhancing global food security. *BIO Web of Conferences, 76*, 05010. https://doi.org/10.1051/bioconf/20237605010

Ambikapathi, R., Schneider, K. R., Davis, B., Herrero, M., Winters, P., & Fanzo, J. C. (2022). Global food systems transitions have enabled affordable diets but had less favourable outcomes for nutrition, environmental health, inclusion and equity. *Nature Food, 3*(9), 764–779.

Anderson, J. A., Nehra, N., Souza, T. L. P. O., Tripathi, L., Layton, R., Hurst, S., Shah, D. M., Gipmans, M., & Pickett, J. (2016). Emerging agricultural biotechnologies for sustainable agriculture and food security. *Journal of Agricultural and Food Chemistry, 64*(2), 383–393. https://doi.org/10.1021/acs.jafc.5b04543

Aruoma, O. I. (2008). Chapter 11—World trade organization and food regulation: Impact on the food supply chain. In *Nutraceutical and functional food regulations in the United States and around the world* (pp. 143–157). Elsevier. https://doi.org/10.1016/b978-012373901-8.00011-1

Atosina Akuriba, M., Alhassan, A. -R., & Abunga Akudugu, M. (2021). *Agri-food and agribusiness systems* (pp. 19–32). springer. https://doi.org/10.1007/978-3-030-88759-9_2

Ayed, R., & Hanana, M. (2021). Artificial intelligence to improve the food and agriculture sector. *Journal of Food Quality, 2021*, 1–7. https://doi.org/10.1155/2021/5584754

Barrett, C. B. (2013). *Food Security and Sociopolitical Stability.* https://doi.org/10.1093/acprof:oso/9780199679362.001.0001

Baune, M.-C., Terjung, N., Tülbek, M. Ç., & Boukid, F. (2022). Textured vegetable proteins (TVP): Future foods standing on their merits as meat alternatives. *Future Foods, 6*, Article 100181. https://doi.org/10.1016/j.fufo.2022.100181

Béné, C., Barange, M., Subramanian, S., Wiebe, K., & Williams, M. (2019). Investing in sustainable food systems. *Nature Sustainability, 2*(5), 392–399. https://doi.org/10.1038/s41893-019-0172-0

Bilali, H., Callenius, C., Strassner, C., & Probst, L. (2018). Food and nutrition security and sustainability transitions in food systems. *Food and Energy Security, 8*(2). https://doi.org/10.1002/fes3.154

Blay-Palmer, A., Conaré, D., Meter, K., Battista, A., & Johnston, C. (2019). *Sustainable Food System Assessment.* https://doi.org/10.4324/9780429439896

Blum, W. E. H. (2013). Soil and land resources for agricultural production: General trends and future scenarios-a worldwide perspective. *International Soil and Water Conservation Research, 1*(3), 1–14. https://doi.org/10.1016/s2095-6339(15)30026-5

Bushueva, A., Adeleye, T., & Roy, P. (2024). Socioeconomic and environmental prospects of the food industry. *Agricultural & Rural Studies, 2*(3), 0016. https://doi.org/10.59978/ar02030016

Chatterjee, R. (2024). Exploring the role of food safety practices in enhancing sustainability within the food industry: A comprehensive analysis of challenges, opportunities, and global trends. *International Journal for Multidimensional Research Perspectives, 2*(9), 56–72. https://doi.org/10.61877/ijmrp.v2i9.189

Dadakas, D., & Tatsi, S. (2021). Global agricultural trade impact of the 2011 triple disaster in Japan: A gravity approach*. *Australian Journal of Agricultural and Resource Economics, 65*(4), 937–972. https://doi.org/10.1111/1467-8489.12448

Davis, K., Downs, S., & Gephart, J. (2020). Towards food supply chain resilience to environmental shocks. *Nature Food, 2*(1), 54–65. https://doi.org/10.1038/s43016-020-00196-3

El Bilali, H., & Ben Hassen, T. (2024). Disrupted harvests: How Ukraine—Russia war influences global food systems—a systematic review. *Policy Studies, 45*(3–4), 310–335. https://doi.org/10.1080/01442872.2024.2329587

Ericksen, P. J. (2008). Conceptualizing food systems for global environmental change research. *Global Environmental Change, 18*(1), 234–245.

Ericksen, P., Stewart, B., Dixon, J., Barling, D., Loring, P., Anderson, M., & Ingram, J. (2012). The value of a food system approach. In *Food security and global environmental change* (pp. 25–45). Routledge.

Falkner, R. (2008). *Agricultural biotechnology* (pp. 140–187). palgrave macmillan uk. https://doi.org/10.1057/9780230277892_5

Fan, S. (2014). *Economics of food insecurity and malnutrition* (pp. 23–31). https://doi.org/10.22004/ag.econ.225569

FAO. (2020). *The state of food and agriculture 2020: Overcoming water challenges in agriculture.* Food and Agriculture Organization of the United Nations.

FAO. (2021). *The impact of COVID-19 on global food systems.* Food and Agriculture Organization of the United Nations.

Garnett, T. (2013). Food sustainability: Problems, perspectives and solutions. *Proceedings of the Nutrition Society, 72*(1), 29–39.

Glavin, M. O. (2003). A single microbial sea: Food safety as a global concern. *SAIS Review, 23*(1), 203–220. https://doi.org/10.1353/sais.2003.0012

Godfray, H. C. J., Beddington, J. R., Crute, I. R., Haddad, L., Lawrence, D., Muir, J. F., & Pretty, J. (2010). Food security: The challenge of feeding 9 billion people. *Science, 327*(5967), 812–818.

Golden, C. D., Gephart, J. A., Eurich, J. G., Mccauley, D. J., Sharp, M. K., Andrew, N. L., & Seto, K. L. (2021). Social-ecological traps link food systems to nutritional outcomes. *Global Food Security, 30*, Article 100561. https://doi.org/10.1016/j.gfs.2021.100561

Gordon, L. J., Falkenmark, M., Folke, C., Jönsson, B. F., Steffen, W., & Johannessen, Å. (2005). Human modification of global water vapor flows from the land surface. *Proceedings of the National Academy of Sciences, 102*(21), 7612–7617. https://doi.org/10.1073/pnas.0500208102

Hakkim, V., Mufeedha, K., Joseph, E., & Gokul, A. (2016). Precision farming: The future of Indian agriculture. *Journal of Applied Biology and Biotechnology, 4*(6), 068–072. https://doi.org/10.7324/jabb.2016.40609

Hertel, T. W. (2018). Food security and international trade. *Annual Review of Resource Economics, 10*, 25–44.

Jagtap, S., Trollman, H., Trollman, F., García-García, G., Parra-López, C., Duong, L., & Afy-Shararah, M. (2022). The russia-ukraine conflict: Its implications for the global food supply chains. *Foods, 11*(14), 2098. https://doi.org/10.3390/foods11142098

Jeswani, H. K., & Azapagic, A. (2022). Environmental sustainability issues in food systems. In *Reference Module in Food Science* (pp. 187–193). Elsevier. https://doi.org/10.1016/b978-0-12-823960-5.00074-3

Jones, A. (2002). An environmental assessment of food supply chains: A case study on dessert apples. *Environmental Management, 30*(4), 560–576. https://doi.org/10.1007/s00267-002-2383-6

Jones, K., Pfeifer, K., & Castillo, G. (2019). Trends in the global food system and implications for institutional foodservice. In *Institutions as conscious food consumers* (pp. 21–46). Academic Press.

Joshi, A., Doriya, R., & Lachure, J. (2023). Drone security for precision agriculture by using one-dimensional convolutional neural network. *Journal of Uncertain Systems, 16*(04). https://doi.org/10.1142/s1752890923500071

Kaphahn, L., & Brennan, L. (2016). Reassessing the risk conditions for political instability in the light of the Arab Spring. *Journal of Risk Research, 20*(11), 1497–1515. https://doi.org/10.1080/13669877.2016.1179211

Khan, A. S. (2017, July 1). *Nanotechnology applications in food and agriculture*. https://doi.org/10.1109/nano.2017.8117471

Kiosia, A., Briend, A., Prendergast, A. J., Dagbasi, A., Morrison, D., Frost, G., Berkley, J. A., Li, J. V., Mathers, J. C., Wilding, J. P. H., Swann, J., Maitland, K., Drake, L., & Kerac, M. (2024). The double burden of malnutrition in individuals: Identifying key challenges and re-thinking research focus. *Nutrition Bulletin, 49*(2). https://doi.org/10.1111/nbu.12670

Kondratowicz, J., Burczyk, E., & Dąbrowska, P. (2009). Żywność genetycznie modyfikowana-szanse i zagrożenia dla konsumentów (część 1). *Chłodnictwo: organ Naczelnej Organizacji Technicznej, 44*(9), 52–57.

Kramkowska, M., Grzelak, T., & Czyzewska, K. (2013). Benefits and risks associated with genetically modified food products. *Annals of Agricultural and Environmental Medicine, 20*(3).

Krishnamoorthy, S., Moses, J. A., & Anandharamakrishnan, C. (2021). COVID-19, food safety, and consumer preferences: changing trends and the way forward. *Journal of Culinary Science & Technology, 21*(5), 719–736. https://doi.org/10.1080/15428052.2021.2016526

Kumar, S., Chandra, A., & Pandey, K. C. (2008). Bacillus thuringiensis (Bt) transgenic crop: An environment friendly insect-pest management strategy. *Journal of Environmental Biology, 29*(5), 641–653.

Kusch-Brandt, S. (2020). Towards more sustainable food systems—14 lessons learned. *International Journal of Environmental Research and Public Health, 17*(11), 4005. https://doi.org/10.3390/ijerph17114005

Lambin, E. F., & Meyfroidt, P. (2011). Global land use change, economic globalization, and the looming land scarcity. *Proceedings of the National Academy of Sciences, 108*(9), 3465–3472. https://doi.org/10.1073/pnas.1100480108

Laso, J., Hoehn, D., Margallo, M., Garcia-Herrero, I., Batlle-Bayer, L., Bala, A., & Aldaco, R. (2018). Assessing energy and environmental efficiency of the spanish agri-food system using the lca/dea methodology. *Energies, 11*(12), 3395. https://doi.org/10.3390/en11123395

Laureati, M., & Proserpio, C. (2024). *Consumer perception and acceptance of healthy and sustainable food: Trends, challenges and mitigatory strategies* (pp. 405–423). Royal Society of Chemistry. https://doi.org/10.1039/bk9781839166655-00405

Lee, A. C. K., Hall, J. A., & Mandeville, K. L. (2011). Global public health training in the UK: Preparing for the future. *Journal of Public Health, 33*(2), 310–316. https://doi.org/10.1093/pubmed/fdr018

Lemos, L. N., Pedrinho, A., Vasconcelos, A. T. R. D., Tsai, S. M., & Mendes, L. W. (2020). Amazon deforestation enriches antibiotic resistance genes. *Soil Biology and Biochemistry, 153*, Article 108110. https://doi.org/10.1016/j.soilbio.2020.108110

Lin, C., Hur, J., Zhang, M., Zhang, Y., Zhang, L., Huang, Z., & Li, W. (2023). *Alleviating heat stress and water scarcity in cultivation with biodegradable radiative cooling mulch*. https://doi.org/10.21203/rs.3.rs-2922832/v1

Lindgren, E., Hiramatsu, M., Gasparatos, A., Loken, B., Haines, A., Javadi, F., Scheelbeek, P., Murakami, T., Harris, F., & Dangour, A. D. (2018). Sustainable food systems\u2014a health perspective. *Sustainability Science, 13*(6), 1505–1517. https://doi.org/10.1007/s11625-018-0586-x

De Lipsis, V., Bleischwitz, R., & Agnolucci, P. (2017). *The international commodity trade* (pp. 297–318). Routledge. https://doi.org/10.4324/9781315560625-20

Liu, X., Liu, W., Tang, Q., Liu, B., Wada, Y., & Yang, H. (2022). Global agricultural water scarcity assessment incorporating blue and green water availability under future climate change. *Earth S Future, 10*(4). https://doi.org/10.1029/2021ef002567

Liu, K., Harrison, M., Yan, H., Liu, D., Meinke, H., Hoogenboom, G., & Zhou, M. (2023). Silver lining to a climate crisis in multiple prospects for alleviating crop waterlogging under future climates. *Nature Communications, 14*(1). https://doi.org/10.1038/s41467-023-36129-4

Lopez-Gunn, E., De Stefano, L., & Llamas, M. R. (2012). The role of ethics in water and food security: Balancing utilitarian and intangible values. *Water Policy, 14*(S1), 89–105. https://doi.org/10.2166/wp.2012.008

Loring, P. A., & Sanyal, P. (2021). Indicators of complexity and over-complexification in global food systems. *Frontiers in Sustainable Food Systems, 5*. https://doi.org/10.3389/fsufs.2021.683100

Lowitt, K., Gray-Donald, K., Hickey, G. M., Francis-Granderson, I., Ville, A. S., Phillip, L. E., & Madramootoo, C. A. (2018). *The obesity pandemic and food insecurity in developing countries* (pp. 255–280). Oxford University Press New York. https://doi.org/10.1093/oso/9780190626686.003.0010

Lucero-Prisno, D. E., Aziato, L., Azanza, P. A., Thepanondh, S., Findlay, I., Huda, A., Dorji, T., Miranda, A. V., Solomon Ethiopia, S., Villaruz, J. F., Adongo, P. B., Li, H., Obnial, J. C., Ogunkola, I. O., Huang, J., Wong, M. C. S., Evardone, S. B., Adebisi, Y. A., Odey, G. O., & Lamawansa, M. D. (2023). Top 10 public health challenges to track in 2023: Shifting focus beyond a global pandemic. *Public Health Challenges, 2*(2). https://doi.org/10.1002/puh2.86

Maertens, M., & Swinnen, J. (2015). *Agricultural trade and development*. https://doi.org/10.30875/c23f13a6-en

Maggio, A., Scapolo, F., van Criekinge, T., & Serraj, R. (2019). Global drivers and megatrends in agri-food systems. In *Agriculture & food systems to 2050: global trends, challenges and opportunities* (pp. 47–83).

Manimekalan, A, & Parthiban, M. S. (2022). Reducing the carbon footprint by cultivating and consuming spirulina: A mini-review. *International Journal of Environment and Climate Change*, 3069–3076. https://doi.org/10.9734/ijecc/2022/v12i111352

Manning, L. (2018). 5—Systems for sustainability and transparency of food supply chains. In *Sustainable food systems from agriculture to industry* (pp. 153–187). Elsevier. https://doi.org/10.1016/b978-0-12-811935-8.00005-6

Maurya, D. K., Chaubey, C., Patel, K. K., Mehta, A. K., Maurya, S. K., Gupta, D., Yadav, R., & Kumar, M. (2024). A review on precision agriculture: An evolution and prospect for the future. *International Journal of Plant & Soil Science, 36*(5), 363–374. https://doi.org/10.9734/ijpss/2024/v36i54534

Mittal, S., Chandel, A. S., & Kumar, S. (2024). *Overcoming barriers to circular supply chain entrepreneurship* (pp. 63–100). IGI Global. https://doi.org/10.4018/979-8-3693-3386-0.ch004

Mondal, B. (2022). *Emerging trends in agriculture: Prospects and opportunities*. RENU. https://doi.org/10.30954/ndp-emerging

Mudzingiri, C., & Chidoko, C. (2014). An investigation of the impact of food aid on food crop production in Zimbabwe. *Mediterranean Journal of Social Sciences, 5*. https://doi.org/10.5901/mjss.2014.v5n27p77

Nadathur, S. R., Wanasundara, J. P. D., & Scanlin, L. (2016). Chapter 1—Proteins in the diet: Challenges in feeding the global population. In *Sustainable protein sources* (pp. 1–19). Elsevier. https://doi.org/10.1016/b978-0-12-802778-3.00001-9

O'Trakoun, J. (2015). Food price uncertainty and political conflict. *International Finance, 18*(3), 299–320. https://doi.org/10.1111/infi.12073

Ochocki, Z. (2005). Stańczak A. Organizmy genetycznie zmodyfikowane–żywność i leki przyszłości? Część I i II. *Farm Pol, 61*(8), 361–377.

Oriekhoe, O., Ikwue, U., Ashiwaju, B., Udeh, C., & Ihemereze, K. (2023). Review of technological advancement in food supply chain management: Comparison between USA and Africa. *World Journal of Advanced Research and Reviews, 20*(3), 1681–1693. https://doi.org/10.30574/wjarr.2023.20.3.2660

Ostaszewska, I., & Sablik, P. (2008). Biotechnologiczne metody doskonalenia zwierząt hodowlanych jako źródła żywności o zwiększonej zawartości kwasów tłuszczowych. *biotechnologia, 3*(82), 153–158.

Pandey, N. (2022). *Agile agriculture supply chain management (AASCM) for managing shifting consumer food preferences: Framework development using grounded theory approach* (pp. 1–12). Springer Nature Singapore. https://doi.org/10.1007/978-981-16-5555-5_1

Phu, N. V. (2021). The impacts of climate change on agricultural production and sustainable agriculture of smallholder farmers in Vietnam. *Vietnam Journal of Agricultural Sciences, 4*(2), 1100–1107. https://doi.org/10.31817/vjas.2021.4.2.10

Pokharel, B. (2023). Advancements in food processing technologies: enhancing safety, quality, and sustainability. *International Journal of Scientific Research in Engineering and Management, 07*(06). https://doi.org/10.55041/ijsrem23682

Porkka, M., Kummu, M., Siebert, S., & Varis, O. (2013). From food insufficiency towards trade dependency: A historical analysis of global food availability. *PLoS ONE, 8*(12), Article e82714. https://doi.org/10.1371/journal.pone.0082714

Reddy, P. B. (2015). Framing gm crops as a solution for global food security. *International Journal of Research—GRANTHAALAYAH, 3*(9SE), 1–5. https://doi.org/10.29121/granthaalayah.v3.i9se.2015.3117

Ruel, M. T., Garrett, J., Yosef, S., & Olivier, M. (2017). Urbanization, food security and nutrition (pp. 705–735). Springer. https://doi.org/10.1007/978-3-319-43739-2_32

Sakakibara, K. Y., & Saito, K. (2006). Genetically modified plants for the promotion of human health. *Biotechnology Letters, 28*, 1983–1991.

Sakapaji, S. (2023). Harnessing ai for climate-resilient agriculture: opportunities and challenges. *European Journal of Theoretical and Applied Sciences, 1*(6), 1144–1158. https://doi.org/10.59324/ejtas.2023.1(6).111

Sanders, C., Mayfield-Smith, K., & Lamm, A. (2021). Exploring twitter discourse around the use of artificial intelligence to advance agricultural sustainability. *Sustainability, 13*(21), 12033. https://doi.org/10.3390/su132112033

Silvestrini, M. M., Smith, N. W., & Sarti, F. M. (2023). Evolution of global food trade network and its effects on population nutritional status. *Current Research in Food Science, 6*, Article 100517. https://doi.org/10.1016/j.crfs.2023.100517

Smith, P., Davis, M., & Roberts, T. (2019). Food systems and climate change: Addressing future challenges. *Global Environmental Change, 59*, 101983.

Smith, N. W., Fletcher, A. J., Hill, J. P., & Mcnabb, W. C. (2021). Animal and plant-sourced nutrition: Complementary not competitive. *Animal Production Science, 62*(8), 701–711. https://doi.org/10.1071/an21235

Soria-Lopez, A., Garcia-Perez, P., Cao, H., Prieto, M. A., Carpena, M., Garcia-Oliveira, P., Fraga-Corral, M., Otero, P., & Simal-Gandara, J. (2022). Challenges for future food systems: From the Green Revolution to food supply chains with a special focus on sustainability. *Food Frontiers, 4*(1), 9–20. https://doi.org/10.1002/fft2.1730-12-820521-1.00003-4

Sousa, R. D., Oliveira, R. S., Da Silva, M. V., & Bragança, L. (2024). Challenges and solutions for sustainable food systems: The potential of home hydroponics. *Sustainability, 16*(2), 817. https://doi.org/10.3390/su16020817

Sylvester, J. M., Bruun, T. B., Castro-Nunez, A., Mertz, O., Gutiérrez-Zapata, D. M., Pérez-Marulanda, L., & Vanegas-Cubillos, M. (2024). *Analysis of food system drivers of deforestation highlights foreign investments and urbanization as threats to tropical forests.* Research Square Platform LLC. https://doi.org/10.21203/rs.3.rs-3901209/v1

Tefft, J., & Jonasova, M. (2020). Food systems transformation in an urbanizing world. In *Handbook on urban food security in the global south* (pp. 34–61). Edward Elgar Publishing.

Trivedi, P., & Sainkhediya, J. (2024). *Precision farming using IoT for smart farming* (pp. 87–93). Bentham Science. https://doi.org/10.2174/9789815223255124010010

Van Berkum, S. (2021). How trade can drive inclusive and sustainable food system outcomes in food deficit low-income countries. *Food Security, 13*(6), 1541–1554. https://doi.org/10.1007/s12571-021-01218-z

Veluru, C. S. (2024). Robotics and data science for smart and precision agriculture. *Journal of Artificial Intelligence & Cloud Computing, 3*(2), 1–7. https://doi.org/10.47363/jaicc/2024(1)306

Vittuari, M., De Menna, F., García-Herrero, L., Pagani, M., Brenes-Peralta, L., & Segrè, A. (2019). Chapter 17—Food systems sustainability: The complex challenge of food loss and waste. In *Sustainable food supply chains* (pp. 249–260). elsevier. https://doi.org/10.1016/b978-0-12-813411-5.00017-x

Wang, J., & Dai, C. (2021). Evolution of global food trade patterns and its implications for food security based on complex network analysis. *Foods, 10*(11), 2657. https://doi.org/10.3390/foods10112657

Wells, J. C. K., Friis, H., Siervo, M., Marphatia, A. A., Miranda, J. J., Haisma, H. H., Amable, G., & Raubenheimer, D. (2021). The future of human malnutrition: Rebalancing agency for better nutritional health. *Globalization and Health, 17*(1). https://doi.org/10.1186/s12992-021-00767-4

White, R., & Yeates, J. (2018). *Farming and climate change* (pp. 315–330). Policy. https://doi.org/10.1332/policypress/9781447336013.003.0020

Wilkinson, J. (2009). The globalization of agribusiness and developing world food systems. *Monthly Review, 61*(4), 29. https://doi.org/10.14452/mr-061-04-2009-08_4

Xu, R., & Li, C. (2022). A modular agricultural robotic system (mars) for precision farming: Concept and implementation. *Journal of Field Robotics, 39*(4), 387–409. https://doi.org/10.1002/rob.22056

Zilberman, D., Holland, T. G., & Trilnick, I. (2018). Agricultural GMOs—what we know and where scientists disagree. *Sustainability, 10*(5), 1514. https://doi.org/10.3390/su10051514

Chapter 12
Food and Water Security and Safety-One Health

Kaviyavarshini Bhavani Shanmugasundaram, Dhanalakshmi Thiyajaraja, Suganthi Varadhaiyan, and Kerenhap Jyotsna Sweety Meshack

Abstract Food safety is an important task in this scenario, and it is based on applying proper hygiene practices along the whole food production chain, from farm to table. The application of realistic, thorough, affordable, and efficient methods such as Hazard Analysis and Critical Control Points, Good Manufacturing Practices, Good Agricultural Practices, the use of lowcost technologies for traceability, frequent inspection and auditing, as well as education and training are necessary to ensure food safety. The challenges of food and water security in an increasingly interconnected world are closely linked to the health of people, animals, and the environment a concept known as the One Health approach. The first section of the text discusses the threats that climate change, population growth, and sociopolitical dynamics pose to the world's food and water security today. It delves deeper into the safety concerns surrounding the production, distribution, use of food, water as well as the depletion of resources. Thus, unsanitary food and water practices will have an impact on food and water systems, leading to pollution, antibiotic resistance, and the emergence of infectious illnesses. These factors will then have an impact on ecosystems, animals, and human health. It does this through a variety of case studies and expert contributions. This Chapter explores the critical concerns surrounding the safety and security of food and water from a One Health standpoint, emphasizing the relationship between public health, agriculture, and ecosystems. It emphasizes how important it is for different stakeholders such as communities, governments, and scientists to work together in order to develop resilience against upcoming challenges.

Keywords Hygienic · Pollution · One health strategy · Food and water security

K. B. Shanmugasundaram (✉) · D. Thiyajaraja · S. Varadhaiyan · K. J. S. Meshack
Nehru Institute of health science, Coimbatore, Tamilnadu, India
e-mail: nihskaviyavarshni@nehrucolleges.com

J. Parray and A. K. Haghi (eds.), *Food and Water Security*,
https://doi.org/10.1007/978-3-032-04870-7_12

12.1 Introduction

Food and water security, along with safety, are vital components of public health and environmental sustainability. Ensuring access to clean, nutritious, and safe food and water not only enhances human health but also strengthens societal resilience. Nevertheless, environmental pollution and climate change are exacerbating threats to these critical resources, creating complex challenges that extend beyond individual fields of study. Industrial and agricultural practices introduce pollutants into water bodies and soils, compromising both food and water safety and resulting in various health and ecological issues. The One Health strategy provides a holistic framework to tackle these interconnected challenges by highlighting the interdependence of human, animal, and environmental health. This approach acknowledges that human health is closely tied to the well-being of animals and ecosystems, promoting collaborative initiatives across different sectors to comprehensively address threats to food and water security. By integrating environmental monitoring, policy development, and public health initiatives, the One Health framework aims to reduce pollution, enhance hygiene, and ensure the availability of safe resources (Cone et al., 1992). Furthermore, achieving food and water security necessitates proactive actions to uphold hygiene standards, which are crucial for preventing disease transmission. Implementing rigorous hygiene protocols in food production, water purification, and resource management is essential for minimizing contamination and alleviating health risks. As populations expand and climate-related challenges grow, the protection of food and water resources becomes increasingly critical. This objective can only be realized through coordinated, interdisciplinary efforts that emphasize pollution management, sustainable practices, and the enhancement of resilience (Rudi et al., 2005).

12.1.1 Threats of Climate Change to Food & Water Security and Safety within the One Health

Climate change represents a significant danger to food and water security, influencing both the availability and quality of these vital resources while also increasing risks to the health of humans, animals, and the environment. Through the lens of the One Health framework, these interrelated issues illustrate the ways in which climate-induced alterations affect the stability of ecosystems that are crucial for ensuring sustainable food and water resources (Adak et al. 2005).

12.1.1.1 Food Security Challenges

Climate change is significantly impacting agricultural productivity across the globe, resulting in reduced food availability and access. Severe weather phenomena such as droughts, floods, and heatwaves harm crops, lower yields, and disrupt food distribution networks. Increasing temperatures and altered precipitation patterns further facilitate the proliferation of pests and plant diseases, which can lead to crop failures and diminished food quality. As agricultural systems face difficulties in adapting, food security for at-risk populations becomes increasingly uncertain (Coffman et al., 1982).

12.1.1.2 Water Scarcity and Quality Issues

The implications of climate change on water resources are profoundly concerning. The melting of glaciers, erratic rainfall patterns, and extended periods of drought are diminishing freshwater supplies in numerous areas. Water scarcity has dire repercussions for agriculture, livestock, and human populations, while deteriorating water quality due to pollution runoff and elevated temperatures promotes the growth of harmful pathogens and toxic algal blooms. This situation not only jeopardizes drinking water sources but also threatens aquatic ecosystems and food safety (Cone et al., 1992).

12.1.1.3 Public Health Consequences

Changes in climate affecting food and water systems are leading to an increase in infectious diseases and malnutrition. Insufficient access to clean water results in inadequate hygiene and sanitation, which promotes the spread of waterborne illnesses such as cholera and dysentery. Furthermore, the risk of foodborne diseases may escalate due to contamination in warmer and more humid climates, where pathogens flourish. For example, elevated water temperatures create conducive environments for bacteria like Vibrio, which can contaminate seafood and pose health risks to consumers (Fig. 12.1) (Rudi et al., 2005).

12.1.1.4 Biodiversity and Ecosystem Disturbance

Climate change is a significant factor in habitat destruction and alterations in species distribution, which adversely affect biodiversity and the resilience of ecosystems. Pollinators, essential for food production, are experiencing population declines due to increasing temperatures and changes in their habitats. Likewise, variations in water temperature and quality endanger aquatic species that are vital for human nutrition and ecological balance. The decline in biodiversity compromises ecosystem functions, rendering food systems more susceptible to the effects of climate change (Rueckert et al. 2005).

Fig. 12.1 Dos and Don'ts to prevent food borne illness

12.1.1.5 Consequences for the One Health Framework

The One Health framework highlights the interdependence of environmental, animal, and human health. Climate change disrupts this equilibrium, affecting ecosystems, wildlife habitats, and agricultural areas. To address these challenges, it is essential to implement collaborative and multi-disciplinary strategies aimed at mitigating climate impacts, enhancing resilience in food and water systems, and prioritizing pollution management and conservation efforts. By adopting adaptive policies, sustainable agricultural methods, and effective water resource management, the One Health approach can play a crucial role in ensuring food and water security amid the challenges posed by climate change (Scallan et al., 2011).

12.1.2 Population Growth, Food & Water Security, and the One Health Approach

Rapid population growth poses considerable challenges to food and water security, as the demand for these vital resources escalates. This increased demand can place significant stress on ecosystems, worsen pollution, and complicate the maintenance of hygiene and safety standards. Within the One Health framework, these challenges are recognized as interconnected, highlighting the health relationships among humans, animals, and the environment (Table 12.1) (Dominski et al., 2024). As the population continues to grow, the need for agricultural expansion and heightened water consumption becomes essential to satisfy rising demands; however, these practices also contribute to environmental degradation. The use of fertilizers, pesticides, and the runoff of wastewater contaminate aquatic ecosystems, adversely affecting water quality and posing threats to both human and wildlife health. Furthermore, the expansion of urban and agricultural areas often disrupts natural ecosystems, leading to a decline in biodiversity and a shift in ecological balance. Such disruptions can facilitate the spread of zoonotic diseases pathogens that can be transmitted between animals and humans thereby jeopardizing public health (Donlan 2002).

Table 12.1 Food & Water Security and Safety within the One Health framework

Aspect	Key issues	One health approach	Examples of actions
Food security	– Access to sufficient, safe, and nutritious food	- sustainable food systems integrating human, animal, and environmental health	– Promoting sustainable agriculture, reducing food waste, and ensuring equitable food distribution
Water security	– Availability of clean and adequate water resources	– Cross-sectoral management of water resources to address contamination and scarcity	– Implementing water conservation programs and improving water purification systems
Food safety	– Contamination and foodborne illnesses	– Collaborative monitoring of pathogens in humans, animals, and the environment	– Enhancing food safety regulations and early warning systems for outbreaks
Water safety	– Pollution, heavy metals, and microbial contamination	– Integrated surveillance systems linking environmental and public health data	– Upgrading wastewater treatment facilities and reducing industrial discharge into water bodies
Zoonotic diseases	– Pathogens transmitted through food or water	– One health surveillance of zoonotic pathogens in food chains and water supplies	– Monitoring antimicrobial resistance in livestock and food industries
Climate change impact	– Stress on food and water systems	– Adaptive strategies addressing changes in ecosystems affecting food and water safety	– Promoting climate-resilient crops and reducing greenhouse gas emissions in food production
Policy and governance	– Lack of integrated regulations	– Interdisciplinary collaboration among health, agriculture, and environmental sectors	– Establishing policies aligning with the WHO'S one health framework
Public awareness	– Limited understanding of food and water risks.	– Community engagement and education on food and water safety within a one health context	– Campaigns on safe food handling and the importance of conserving water resources

As the population increases, upholding hygienic standards for food and water becomes more difficult due to the shared nature of limited resources. Population pressures frequently result in overcrowded living conditions, hindering the ability to maintain proper sanitation and hygiene. Inadequate waste disposal and insufficient water treatment facilities can further intensify pollution, resulting in contaminated food and water supplies that threaten health (Dunne 2002). The One Health strategy presents a comprehensive framework for tackling these issues. It highlights the importance of coordinated efforts among the health, environmental, and agricultural sectors to achieve sustainable resource management, mitigate pollution, and safeguard biodiversity. By advocating for sustainable farming methods, optimal water utilization, and rigorous hygiene protocols, the One Health approach seeks to guarantee safe food and water supplies for increasing populations while also safeguarding environmental and animal health. This strategy encourages interdisciplinary collaboration, enhancing resilience and ensuring that both present and future generations can access safe, hygienic, and pollution free food and water resources (Fig. 12.2) (Rudi et al., 2005).

12.1.3 *The Impact of Sociopolitical Dynamics on Global Food and Water Security*

Sociopolitical dynamics are essential in influencing the global landscape of food and water security. Factors such as geopolitical conflicts, unequal distribution of resources, governance issues, and economic inequalities significantly affect the accessibility and safety of food and water. These challenges often hinder the ability to uphold sanitary standards and mitigate pollution. The One Health approach, which encompasses human, animal, and environmental health, underscores the importance of collaborative initiatives across various sectors to tackle the urgent issues related to food and water security (Rueckert 2005; Watson, et al. 2013).

12.1.3.1 Resource Conflicts and Access Inequality

In areas where water and cultivable land are limited, sociopolitical tensions often emerge regarding the distribution of resources. These conflicts can severely disrupt food production and access to water, particularly for marginalized groups. Political instability frequently results in insufficient regulatory measures, which can compromise water quality and agricultural safety, thereby obstructing the adoption of sanitary practices and pollution management. For instance, in areas affected by conflict, water sources may be redirected or contaminated, adversely impacting both human and animal health and jeopardizing food and water security (Scallan, et al., 2011; Kiaya et al., 2014);(Food and Agriculture Organization of the United Nations (FAO) 2018).

Fig. 12.2 One Health and Water

12.1.3.2 Trade Policies and Economic Disparities

Global trade regulations and economic inequalities can hinder the availability of safe and nutritious food, especially in low-income areas. Trade barriers or sanctions can disrupt food supply chains, resulting in shortages or inflated prices that make nutritious food inaccessible to many communities. Furthermore, lower-income nations often lack the necessary infrastructure and resources to uphold hygienic standards in food production and water treatment, rendering them susceptible to pollution and contamination. Disparities in resource access also impede the effectiveness of the One Health approach, as economically disadvantaged regions face challenges in maintaining food and water safety standards (Scallan et al., 2011; Stewart et al., 1997). (Stirling, 2003).

12.1.3.3 Environmental Policies and Industrial Practices

Inadequate or inconsistent environmental regulations worsen pollution, adversely affecting food and water safety. Industrial operations, such as mining and chemical manufacturing, can contaminate water sources and agricultural land if not properly regulated, threatening the health of nearby communities and ecosystems. Political advocacy for stronger environmental protections, pollution management, and sanitary practices in agriculture and water management is crucial for establishing a sustainable food and water security framework. The One Health paradigm promotes policies that consider the environmental effects on food and water systems to safeguard human and animal health from pollution (Tholozan et al., 1999).

12.1.4 Safety Concerns in Food and Water Security

Food and water security are critical pillars of global health and sustainability, closely tied to the principles of hygiene, pollution control, and the One Health strategy. Ensuring hygienic practices across the production, distribution, and use of food and water is paramount to prevent contamination and the spread of diseases. Pollution from industrial, agricultural, and domestic activities poses a severe threat to the safety of these resources, compromising their quality and availability. Depletion of water resources and arable land due to unsustainable practices further exacerbates the challenge of securing these essentials for future generations. The One Health approach underscores the interconnectedness of human, animal, and environmental health, emphasizing the need for integrated strategies to mitigate risks. Through pollution control, sustainable resource management, and promotion of hygienic standards, we can strengthen food and water security while aligning with broader One Health objectives to safeguard global well-being (Wang et al., 1997).

12.1.4.1 Hygienic Concerns

Food Production and Handling: Contamination at various stages from farming, processing, storage, to distribution poses risks. Poor hygiene can introduce pathogens like Salmonella, E. coli, or Listeria (Lillford et al., 2021).

Water Sources: Unsafe practices, including the improper treatment of water, lead to the spread of waterborne diseases like cholera, typhoid, and diarrhea.

Personal Hygiene: Improper food preparation, lack of handwashing, and unhygienic kitchen environments exacerbate risks in households.

12.1.4.2 Pollution

(a) Water Pollution:

Chemical Runoff: Fertilizers, pesticides, and industrial discharges contaminate water supplies with nitrates, heavy metals, and other toxins.

Microplastics: Their presence in water sources threatens both ecosystem health and human safety.

Sewage Overflow: Inadequate sanitation systems can introduce pathogens and pollutants into drinking water.

(b) Food Contamination:

Soil Degradation: Pollutants in the soil affect crop safety and yield.

Airborne Pollutants: Deposition of contaminants like dioxins or heavy metals during food production.

12.1.4.3 Resource Depletion

Overuse of Water Resources: Agriculture accounts for 70% of global freshwater use, often unsustainably. Excessive groundwater withdrawal leads to aquifer depletion.

Soil Erosion and Loss of Fertility: Unsustainable agricultural practices reduce soil's ability to support healthy crops.

Climate Change: Rising temperatures and erratic weather patterns affect food and water availability.

12.1.5 Role of the One Health Strategy

The One Health Strategy emphasizes the interconnectedness of human, animal, and environmental health in addressing food and water security challenges:

Integrated Surveillance Systems: Early detection of zoonotic diseases (e.g., Salmonella or Campylobacter from animals) prevents outbreaks linked to food and water.

Antimicrobial Resistance (AMR): Overuse of antibiotics in agriculture contributes to AMR, posing risks to public health. One Health promotes prudent use and monitoring.

Environmental Protection: Mitigating pollution and managing ecosystems (like wetlands) ensures safer water and sustainable farming (Vass et al., 2009).

12.1.6 Potential Solutions and Preventive Measures

Hygiene and Safety Standards: Enforcing Good Agricultural Practices (GAPs) and Hazard Analysis Critical Control Points (HACCP) in food production. Expanding access to clean water and sanitation facilities (Adoff et al., 2015; Walton, 2013; Wang, 2019).

1. **Pollution Control:** Regulating industrial discharges and adopting eco-friendly farming practices. Implementing wastewater treatment and safe reuse policies.
2. **Sustainable Resource Management:** Promoting efficient irrigation methods like drip irrigation. Encouraging crop rotation and soil conservation techniques.
3. **Technological Innovations:** Utilizing blockchain for traceability in food supply chains. Deploying biosensors for real-time contamination detection in water and food (The World Development Bank. World Development Report 2020 UNICEF Annual Report 2015).

12.1.7 Conclusion

This chapter underscores the pressing need to address the safety and security of food and water through a One Health lens, recognizing the interconnectedness of public health, agriculture, and ecosystems. Ensuring hygienic practices and mitigating pollution are pivotal in safeguarding these critical resources against contamination and degradation. The chapter highlights that collaborative efforts among communities, governments, scientists, and other stakeholders are essential to develop sustainable strategies for food and water security. By adopting the One Health strategy, we can build resilience against emerging challenges, ensuring the health of humans, animals, and the environment while securing vital resources for future generations.

References

Adak, G. K., Meakins, S. M., Yip, H., Lopman, B. A., & O'Brien, S. J. (2005). Disease risks from foods, England and Wales, 1996–2000. *Emerging Infectious Disease, 11*, 365–372.

Coffman, G. L., Gaubatz, J. W., Yielding, K. L., & Yielding, L. W. (1982). Demonstration of specifc high affnity binding sites in plasmid DNA by photoaffnity labelling with ethidium analogue. *Journal of Biological Chemistry, 257*, 13205–13297.

Cone, R. W., Hobson, A. C., & Huang, M. L. (1992). Coamplifed positive control detects inhibition of polymerase chain reactions. *Journal of Clinical Microbiology, 30*, 3185–3189.

Dominski, B. H., Raitz, M. F., Provenzi, M. A., de Oliveira Silveira, A. C., Marques Sincero, T. C., & Antunes Ferreira, F. (2024). Characterization of methicillin-resistant Staphylococcus aureus (MRSA) isolated in Santa Catarina (SC), Brazil. *Diagnostic Microbiology and Infectious Disease, 109*(1), 16244. https://doi.org/10.1016/j.diagmicrobio.2024.116244

Donlan, R. M. (2002). Biofilms: Microbial life on surfaces. *Emerging Infectious Diseases, 8*(9), 881–890. https://doi.org/10.3201/eid0809.020063

Dunne, W. M. (2002). Bacterial adhesion: Seen any good biofilms lately? *Clinical Microbiology Reviews, 15*(2), 155–166. https://doi.org/10.1128/CMR.15.2.155-166.2002

Food and Agriculture Organization of the United Nations (FAO). 2018. Transforming food and agriculture to achieve the sdgs. 20 interconnected actions to guide decision-makers; Food and Agriculture Organization of the United Nations: Rome, Italy. http://www.fao.org/3/I9900EN/i9900en.pdf (accessed on 3 June 2021).

Food and Agriculture Organization of the United Nations (FAO). 2019. How to reduce food loss and waste for food security and environmental sustainability; FAO: Rome, Italy. http://www.fao.org/3/ca6481en/ca6481en.pdf (accessed on 2 July 2021).
Kiaya, V. (2014). Post-harvest losses and strategies to reduce them. technical paper on post-harvest losses. Action contre la faim. https://www.actioncontrelafaim.org/wp-content/uploads/2018/01/technical_paper_phl__.pdf (accessed on 3 June 2021).
Lillford, P., & Hermansson, A. M. (2021). Global missions and the critical needs of food science and technology. *Trends in Food Science & Technology, 111*, 800–811. [CrossRef].
Rudi, K., Moen, B., Dromtorp, S. M., & Holck, A. L. (2005). Use of ethidium monoazide and PCR in combination for quantifcation of viable and dead cells in complex samples. *Applied and Environmental Microbiology, 71*, 1018–1024.
Rueckert, A., Ronimus, R. S., & Morgan, H. W. (2005). Rapid differentiation and enumeration of the total, viable vegetative cell and spore content of thermophilic bacilli in milk powders with reference to Anoxybacillus favithermus. *Journal of Applied Microbiology, 99*, 1246–1255.
Sadoff, C. W., Hall, J. W., Grey, D., Aerts, J. C. J. H., Ait-kadi, M., Brown, C., et al. (2015). *Securing water, sustaining growth: Report of the GWD/OECD task force on water security and sustainable growth.* University of Oxford UK.
Scallan, E., Griffn, P. M., Angulo, F. J., Tauxe, R. V., & Hoekstra, R. M. (2011c). Foodborne illness acquired in the United States—Unspecifed agents. *Emerging Infectious Diseases, 17*, 16–22.
Scallan, E., Hoekstra, R. M., Angulo, F. J., Tauxe, R. V., Widdowson, M. A., Roy, S. L., Jones, J. L., & Griffn, P. M. (2011). Foodborne illness acquired in the United States—Major pathogens. *Emerging Infectious Diseases, 17*, 7–15.
Stewart, G. S. (1997). Challenging food microbiology from a molecular perspective. *Microbiology, 143*, 2099–2108.
Stirling, D. (2003). Quality control in PCR. *Methods Molecular Biology, 226*, 21–24.
The World Development Bank. (2020). World Development Report 2020 Chapters and Data. http://www.worldbank.org.
Tholozan, J. L., Cappelier, J. M., Tissier, J. P., Delattre, G., & Federighi, M. (1999). Physiological characterization of viable-but-nonculturable campylobacter jejuni cells. *Applied Environmental Microbiology, 65*, 1110–1116.
UNICEF Annual Report 2015. https://www.unicef.org.
Vass, K. K., Das, M. K., Srivastava, P. K., & Dey, S. (2009). Assessing the impact of climate change on inland fisheries in river ganga and its plains in India. *Aquatic Ecosystem Health and Management, 12*(2).
Walton, B. (2013) Report: UN Proposes a Common Definition for Water Security on World Water Day.
Wang, S., & Levin, R. E. (2006). Discrimination of viable vibrio vulnifcus cells from dead cells in real-time PCR. *Journal of Microbiology Methods, 64*, 1–8.
Wang, T. (2014). Global water with drawl and consumption 2014- 2040. Energy and Environment Services. Statista.
Waring, M. J. (1965). Complex formation between ethidium bromide and nucleic acids. *Journal of Molecular Biology, 13*, 269–282.
Watson, M., & Meah, A. (2013) Food, waste and safety: Negotiating conflicting social anxieties into the practices of domestic provisioning. *The Sociological Review, 60* (Suppl. 2), 102–120. [CrossRef].
Wilson, I. G. (1997). Inhibition and facilitation of nucleic acid amplifcation. *Applied Environmental Microbiology, 63*, 3741–3751.
Water Resources Group. (2009). Charting our water future: economic framework to inform decision making.

MIX
Papier aus verantwortungsvollen Quellen
Paper from responsible sources
FSC® C105338

If you have any concerns about our products,
you can contact us on
ProductSafety@springernature.com

In case Publisher is established outside the EU,
the EU authorized representative is:
Springer Nature Customer Service Center GmbH
Europaplatz 3, 69115 Heidelberg, Germany

Printed by Libri Plureos GmbH
in Hamburg, Germany